2024 IPSDC®

INTERNATIONAL **PRIVATE SEWAGE DISPOSAL** CODE®

2024 International Private Sewage Disposal Code®

First Printing: July 2023

ISBN: 978-1-959851-82-0 (soft-cover edition)
ISBN: 978-1-959851-83-7 (PDF download)

T029211

PRINTED IN THE USA

NEW DESIGN FOR THE 2024 INTERNATIONAL CODES

The 2024 International Codes® (I-Codes®) have undergone substantial formatting changes as part of the digital transformation strategy of the International Code Council® (ICC®) to improve the user experience. The resulting product better aligns the print and PDF versions of the I-Codes with the ICC's Digital Codes® content.

The changes, promoting a cleaner, more modern look and enhancing readability and sustainability, include:

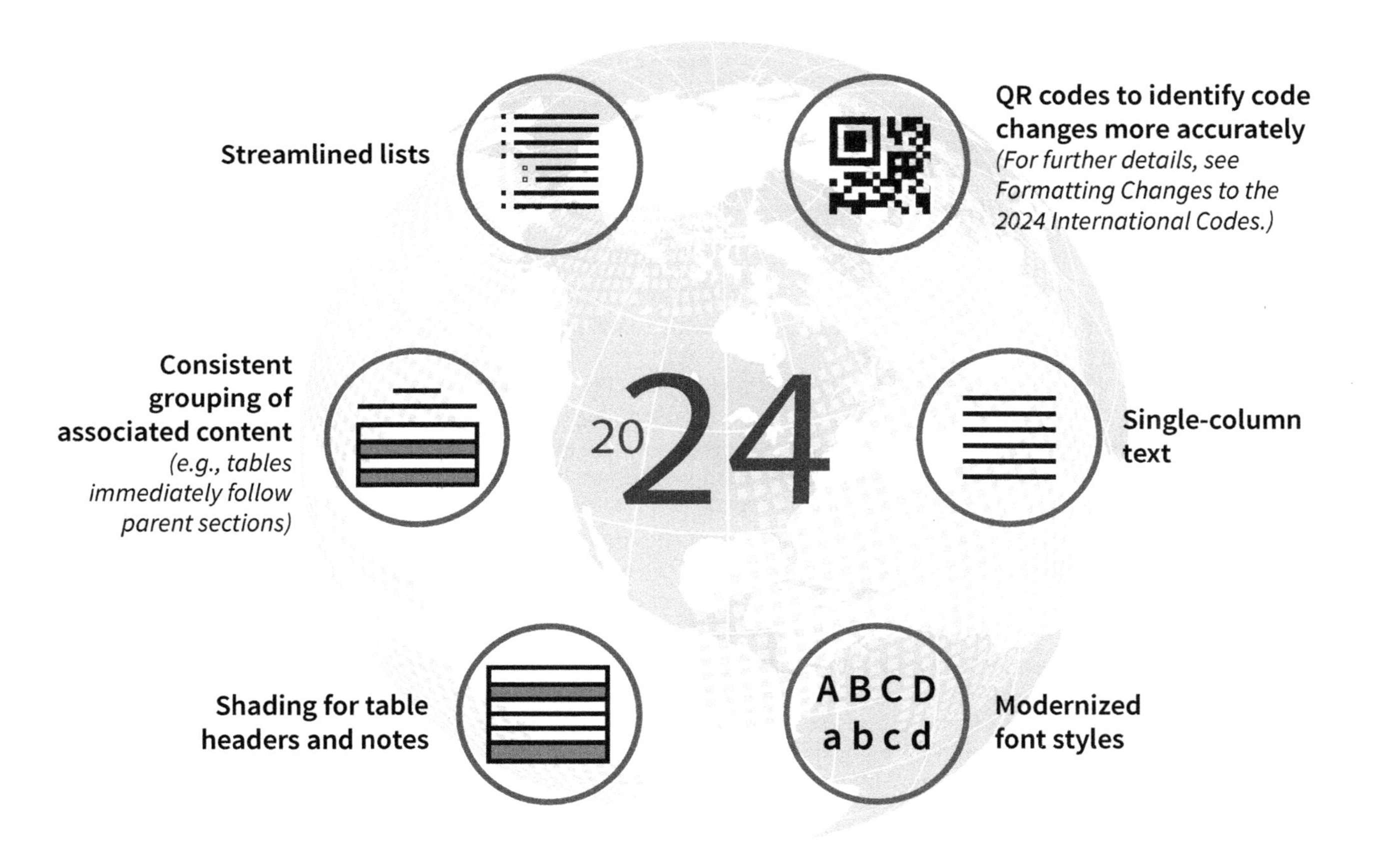

More information can be found at iccsafe.org/design-updates.

PREFACE

FORMATTING CHANGES TO THE 2024 INTERNATIONAL CODES

The 2024 International Codes® (I-Codes®) have undergone substantial formatting changes as part of the digital transformation strategy of the International Code Council® (ICC®) to improve the user experience. The resulting product better aligns the print and PDF versions of the I-Codes with the ICC's Digital Code content. Additional information can be found at iccsafe.org/design-updates.

Replacement of Marginal Markings with QR Codes

Through 2021, print editions of the I-Codes identified technical changes from prior code cycles with marginal markings [solid vertical lines for new text, deletion arrows (➡), asterisks for relocations (*)]. The 2024 I-Code print editions replace the marginal markings with QR codes to identify code changes more precisely.

A QR code is placed at the beginning of any section that has undergone technical revision. If there is no QR code, there are no technical changes to that section.

In the following example from the 2024 *International Private Sewage Disposal Code®* (IPSDC®), a QR code indicates there are changes to Section 109 from the 2021 IPSDC. Note that the change may occur in the main section or in one or more subsections of the main section.

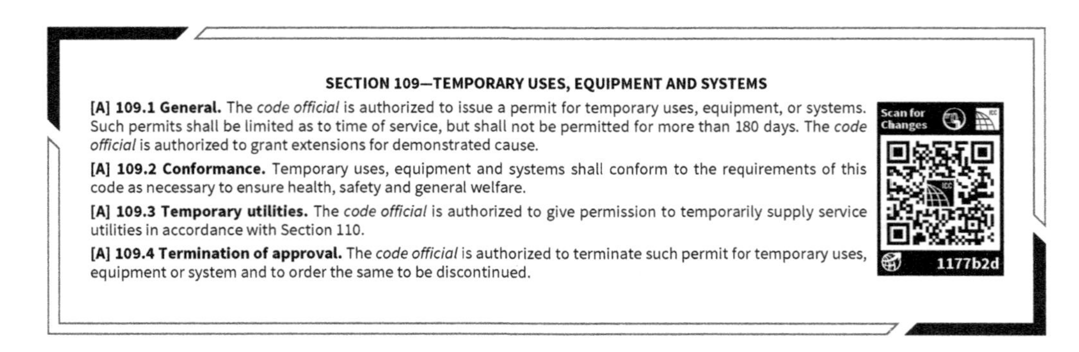

SECTION 109—TEMPORARY USES, EQUIPMENT AND SYSTEMS

[A] 109.1 General. The *code official* is authorized to issue a permit for temporary uses, equipment, or systems. Such permits shall be limited as to time of service, but shall not be permitted for more than 180 days. The *code official* is authorized to grant extensions for demonstrated cause.

[A] 109.2 Conformance. Temporary uses, equipment and systems shall conform to the requirements of this code as necessary to ensure health, safety and general welfare.

[A] 109.3 Temporary utilities. The *code official* is authorized to give permission to temporarily supply service utilities in accordance with Section 110.

[A] 109.4 Termination of approval. The *code official* is authorized to terminate such permit for temporary uses, equipment or system and to order the same to be discontinued.

Scan for Changes

1177b2d

To see the code changes, the user need only scan the QR code with a smart device. If scanning a QR code is not an option, changes can be accessed by entering the 7-digit code beneath the QR code at the end of the following URL: qr.iccsafe.org/ (in the above example, "qr.iccsafe.org/1177b2d"). Those viewing the code book via PDF can click on the QR code.

All methods take the user to the appropriate section on ICC's Digital Codes website, where technical changes from the prior cycle can be viewed. Digital Codes Premium subscribers who are logged in will be automatically directed to the Premium view. All other users will be directed to the Digital Codes Basic free view. Both views show new code language in blue text along with deletion arrows for deleted text and relocation markers for relocated text.

Digital Codes Premium offers additional ways to enhance code compliance research, including revision histories, commentary by code experts and an advanced search function. A full list of features can be found at codes.iccsafe.org/premium-features.

ACCESSING ADDITIONAL FEATURES VIA REGISTRATION OF BOOK

Beginning with the 2024 *International Mechanical Code®* (IMC®) and the 2024 *International Plumbing Code®* (IPC®), users will be able to validate the authenticity of their book and register it with the ICC to receive incentives. Digital Codes Premium (codes.iccsafe.org) provides advanced features and exclusive content to enhance code compliance. To validate and register, the user will tap the ICC tag (pictured here and located on the front cover) with a near-field communication (NFC) compatible device. Visit iccsafe.org/nfc for more information and troubleshooting tips regarding NFC tag technology.

ABOUT THE I-CODES

The 2024 I-Codes, published by the ICC, are 15 fully compatible titles intended to establish provisions that adequately protect public health, safety and welfare; that do not unnecessarily increase construction costs; that do not restrict the use of new materials, products or methods of construction; and that do not give preferential treatment to particular types or classes of materials, products or methods of construction.

The I-Codes are updated on a 3-year cycle to allow for new construction methods and technologies to be incorporated into the codes. Alternative materials, designs and methods not specifically addressed in the I-Code can be approved by the building official where the proposed materials, designs or methods comply with the intent of the provisions of the code.

The I-Codes are used as the basis of laws and regulations in communities across the US and in other countries. They are also used in a variety of nonregulatory settings, including:

- Voluntary compliance programs.
- The insurance industry.

- Certification and credentialing for building design, construction and safety professionals.
- Certification of building and construction-related products.
- Facilities management.
- "Best practices" benchmarks for designers and builders.
- College, university and professional school textbooks and curricula.
- Reference works related to building design and construction.

Code Development Process

The code development process regularly provides an international forum for building professionals to discuss requirements for building design, construction methods, safety, performance, technological advances and new products. Proposed changes to the I-Codes, submitted by code enforcement officials, industry representatives, design professionals and other interested parties are deliberated through an open code development process in which all interested and affected parties may participate.

Openness, transparency, balance, due process and consensus are the guiding principles of both the ICC Code Development Process and OMB Circular A-119, which governs the federal government's use of private-sector standards. The ICC process is open to anyone without cost. Remote participation is available through cdpAccess®, the ICC's cloud-based app.

In order to ensure that organizations with a direct and material interest in the codes have a voice in the process, the ICC has developed partnerships with key industry segments that support the ICC's important public safety mission. Some code development committee members were nominated by the following industry partners and approved by the ICC Board:

- American Gas Association (AGA)
- American Institute of Architects (AIA)
- American Society of Plumbing Engineers (ASPE)
- International Association of Fire Chiefs (IAFC)
- National Association of Home Builders (NAHB)
- National Association of State Fire Marshals (NASFM)
- National Council of Structural Engineers Association (NCSEA)
- National Multifamily Housing Council (NMHC)
- Plumbing Heating and Cooling Contractors (PHCC)
- Pool and Hot Tub Alliance (PHTA), formerly The Association of Pool and Spa Professionals (APSP)

Code development committees evaluate and make recommendations regarding proposed changes to the codes. Their recommendations are then subject to public comment and council-wide votes. The ICC's governmental members—public safety officials who have no financial or business interest in the outcome—cast the final votes on proposed changes.

The I-Codes are subject to change through future code development cycles and by any governmental entity that enacts the code into law. For more information regarding the code development process, contact the Codes and Standards Development Department of the ICC at iccsafe.org/products-and-services/i-codes/code-development/.

While the I-Code development procedure is thorough and comprehensive, the ICC, its members and those participating in the development of the codes expressly disclaim any liability resulting from the publication or use of the I-Codes, or from compliance or noncompliance with their provisions. NO WARRANTY OF ANY KIND, IMPLIED, EXPRESSED OR STATUTORY, IS GIVEN WITH RESPECT TO THE I-CODES. The ICC does not have the power or authority to police or enforce compliance with the contents of the I-Codes.

Code Development Committee Responsibilities
(Letter Designations in Front of Section Numbers)

In each cycle, proposed changes are considered by the Code Development Committee assigned to a specific code or subject matter. Committee Action Hearings result in recommendations regarding a proposal to the voting membership. Where changes to a code section are not considered by that code's own committee, the code section is preceded by a bracketed letter designation identifying a different committee. Bracketed letter designations for the I-Code committees are:

[A] = Administrative Code Development Committee
[BE] = IBC—Egress Code Development Committee
[BF] = IBC—Fire Safety Code Development Committee
[BG] = IBC—General Code Development Committee
[BS] = IBC—Structural Code Development Committee
[E] = Developed under the ICC's Standard Development Process
[EB] = International Existing Building Code Development Committee
[F] = International Fire Code Development Committee
[FG] = International Fuel Gas Code Development Committee

[M] = International Mechanical Code Development Committee

[P] = International Plumbing Code Development Committee

[SP] = International Swimming Pool and Spa Code Development Committee

For the development of the 2027 edition of the I-Codes, the ICC Board of Directors approved a standing motion from the Board Committee on the Long-Term Code Development Process to revise the code development cycle to incorporate two committee action hearings for each code group. This change expands the current process from two independent 1-year cycles to a single continuous 3-year cycle. There will be two groups of code development committees and they will meet in separate years. The current groups will be reworked. With the energy provisions of the *International Energy Conservation Code*® (IECC®) and *Chapter 11* of the *International Residential Code*® (IRC®) now moved to the Code Council's Standards Development Process, the reduced volume of code changes will be distributed between Groups A and B.

Code change proposals submitted for code sections that have a letter designation in front of them will be heard by the respective committee responsible for such code sections. Because different committees hold Committee Action Hearings in different years, proposals for most codes will be heard by committees in both the 2024 (Group A) and the 2025 (Group B) code development cycles. It is very important that anyone submitting code change proposals understands which code development committee is responsible for the section of the code that is the subject of the code change proposal.

Please visit the ICC website at iccsafe.org/products-and-services/i-codes/code-development/current-code-development-cycle for further information on the Code Development Committee responsibilities as it becomes available.

Coordination of the I-Codes

The coordination of technical provisions allows the I-Codes to be used as a complete set of complementary documents. Individual codes can also be used in subsets or as stand-alone documents. Some technical provisions that are relevant to more than one subject area are duplicated in multiple model codes.

Italicized Terms

Words and terms defined in Chapter 2, Definitions, are italicized where they appear in code text and the Chapter 2 definitions apply. Although care has been taken to ensure applicable terms are italicized, there may be instances where a defined term has not been italicized or where a term is italicized but the definition found in Chapter 2 is not applicable. For example, Chapter 2 of the *International Building Code*® (IBC®) contains a definition for "*Listed*" that is applicable to equipment, products and services. The term "listed" is also used in that code to refer to a list of items within the code or within a referenced document. For the latter, the Chapter 2 definition would not be applicable.

Adoption of International Code Council Codes and Standards

The International Code Council maintains a copyright in all of its codes and standards. Maintaining copyright allows the Code Council to fund its mission through sales of books in both print and digital formats. The Code Council welcomes incorporation by reference of its codes and standards by jurisdictions that recognize and acknowledge the Code Council's copyright in the codes and standards, and further acknowledge the substantial shared value of the public/private partnership for code development between jurisdictions and the Code Council. By making its codes and standards available for incorporation by reference, the Code Council does not waive its copyright in its codes and standards.

The Code Council's codes and standards may only be adopted by incorporation by reference in an ordinance passed by the governing body of the jurisdiction. "Incorporation by reference" means that in the adopting ordinance, the governing body cites only the title, edition, relevant sections or subsections (where applicable), and publishing information of the model code or standard, and the actual text of the model code or standard is not included in the ordinance (see graphic, "Adoption of International Code Council Codes and Standards"). The Code Council does not consent to the reproduction of the text of its codes or standards in any ordinance. If the governing body enacts any changes, only the text of those changes or amendments may be included in the ordinance.

ADOPTION OF INTERNATIONAL CODE COUNCIL CODES AND STANDARDS

INCORPORATED BY REFERENCE

What does "incorporate by reference" mean? If a governmental agency or authority having jurisdiction (AHJ) over code adoption wishes to adopt a model code for legislative or regulatory purposes, it will enact an ordinance, regulation or law to incorporate by reference (IBR) the relevant code. The actual text of the model code is not included in the law, but the enacting law will include the full text of any changes or amendments enacted by the legislative body of the AHJ.

Step 1
The International Code Council updates the current model codes every three years.

Step 2
The International Code Council publishes the code both in print and online.

Step 3
The model code review process begins with governments.

Step 4
The adoption timeframe is set.

Step 5
Language is modified and amendments are added or recommended through the review process.

Step 6
AHJs set effective date, and notify stakeholders working on projects.

Step 7
Code is adopted into law through legislative or regulatory action, citing the title, edition and publishing information of the model code.

The Code Council also recognizes the need for jurisdictions to make laws accessible to the public. Accordingly, all I-Codes and I-Standards, along with the laws of many jurisdictions, are available to view for free at codes.iccsafe.org/codes/i-codes.

These documents may also be purchased, in both digital and print versions, at shop.iccsafe.org. To facilitate adoption, some I-Code sections contain blanks for fill-in information that needs to be supplied by the adopting jurisdiction as part of the adoption legislation. For example, the IPSDC contains:

Section 101.1. Insert: **[NAME OF JURISDICTION]**

Section 113.4. Insert: **[OFFENSE, DOLLAR AMOUNT, NUMBER OF DAYS]**

Section 405.2.5. Insert: **[DATE IN THREE LOCATIONS]**

Section 405.2.6. Insert: **[DATE IN TWO LOCATIONS]**

For further information or assistance with adoption, including a sample ordinance, jurisdictions should contact the Code Council at incorporation@iccsafe.org.

For a list of frequently asked questions (FAQs) addressing a range of foundational topics about the adoption of model codes by jurisdictions and to learn more about the Code Council's code adoption resources, scan the QR code or visit iccsafe.org/code-adoption-resources.

INTRODUCTION TO THE INTERNATIONAL PRIVATE SEWAGE DISPOSAL CODE

The *International Private Sewage Disposal Code*® (IPSDC®) establishes minimum requirements for sewage disposal systems using prescriptive and performance-related provisions. It is founded on broad-based principles that make possible the use of new materials and new sewage disposal designs.

The IPSDC is a model code that regulates minimum requirements for the installation of new or the alteration of existing private sewage disposal systems. Where a building cannot be served by a public sewer system, the building site must be provided with a system for treating the wastewater generated from the use of plumbing fixtures in the building. The IPSDC addresses site evaluations, materials, various soil absorption systems, holding tanks, cesspools and on-site wastewater treatment systems. The IPSDC provides a total approach for the on-site, safe disposal of the waste flow discharged to the plumbing fixtures in a building.

The IPSDC is a specification- (prescriptive-) oriented code with very few occurrences of performance-oriented text. The site soil must be evaluated in a prescribed manner to determine its ability to accept the waste flow. The chosen waste treatment method must be designed in a prescribed manner for the soil conditions at the building site, constructed using prescribed materials and installed according to prescribed dimensions. The IPSDC sets forth the minimum acceptable requirements for private sewage disposal systems in order to protect humans and the environment from insanitary conditions that would develop if waste flows were not rendered harmless.

ARRANGEMENT AND FORMAT OF THE 2024 IPSDC

The format of the IPSDC allows each chapter to be devoted to a particular subject with the exception of Chapter 3, which contain subject matters that are not extensive enough to warrant their own independent chapter. The following table shows how the IPSDCis divided. The chapter synopses detail the scope and intent of the provisions of the IPSDC.

CHAPTER TOPICS	
CHAPTERS	**SUBJECTS**
1	Scope and Administration
2	Definitions
3	General Regulations
4	Site Evaluation and Requirements
5	Materials
6	Soil Absorption Systems
7	Pressure Distribution Systems
8	Tanks
9	Mound Systems
10	Cesspools
11	Residential Wastewater Systems
12	Inspections
13	Nonliquid Saturated Treatment Systems
14	Referenced Standards
Appendix A	System Layout Illustrations
Appendix B	Tables for Pressure Distribution Systems
Appendix C	Board of Appeals

Chapter 1 Scope and Administration.

Chapter 1 establishes the limits of applicability of the code and describes how the code is to be applied and enforced. The provisions of Chapter 1 establish the authority and duties of the code official appointed by the authority having jurisdiction and also establish the rights and privileges of the design professional, contractor and property owner.

Chapter 2 Definitions.

Chapter 2 is the repository of the definitions of terms used in the body of the code. The user of the code should be familiar with and consult this chapter because the definitions are essential to the correct interpretation of the code and because the user may not be aware that a term is defined.

Chapter 3 General Regulations.

The content of Chapter 3 is often referred to as "miscellaneous," rather than general regulations. Chapter 3 received that label because it is the only chapter in the code where requirements do not interrelate. If a requirement cannot be located in another chapter, it can be found in this chapter. Specific requirements concerning flood hazard areas are in this chapter.

Chapter 4 Site Evaluation and Requirements.

A private sewage disposal system has an effluent that cannot be directly discharged into waterways or open ponds. Soil of the right consistency and water content provides natural filtering and treatment of this discharge. Because soil conditions vary widely, even on the same building site, tests and inspections of the soils must be performed to evaluate the degree to which the soil can accept these liquids. The results of the tests provide necessary information to design an adequate private sewage disposal system. Chapter 4 provides the methods for evaluating the building site.

Chapter 5 Materials.

Private sewage disposal systems depend on the strength, quality and chemical resistance of the components that make up the system. To that end, the purpose of Chapter 5 is to specify the minimum material and component standards to ensure that the private sewage disposal system will correctly perform for its intended life.

Chapter 6 Soil Absorption Systems.

The design of soil absorption systems depends heavily on the result of the tests and evaluation of the site soil conditions required in Chapter 4. Where soil is less permeable, the area of the soil absorption must be large as compared to that required for soils that are highly permeable. The type of building that is being served by the private sewage disposal system also affects the size of the planned soil absorption area. Chapter 6 provides the methods for computing the required absorption area and details for the proper installation of the soil absorption systems.

Chapter 7 Pressure Distribution Systems.

Chapter 6 deals with gravity-type soil absorption systems or systems where the effluent is allowed to drain out of the distribution piping by gravity. Chapter 7 offers an alternative method of discharging the effluent into the ground by pressure means. As such, Chapter 7 provides the necessary details for designing the piping and pumping systems for pressure distribution systems.

Chapter 8 Tanks.

Tanks are an integral part of any private sewage disposal system, whether they serve as treatment (septic) tanks or merely just holding tanks for leveling the peaks in flow to the system. Where tanks are used for treatment, the dimensions, volume and location of internal features are very important to ensure that the solid wastes are kept within the tank so as to not clog the effluent distribution system. Where tanks are used for holding purposes, they must be sized large enough to accommodate the total of peak flows coming from a building. Chapter 8 provides the necessary requirements for tanks.

Chapter 9 Mound Systems.

Mound systems are another method for applying the effluent from a private sewage disposal system to the soil. This type of system may be advantageous in some localities due to the existing soil conditions. Chapter 9 has specific requirements for soil and site evaluations for mound systems.

Chapter 10 Cesspools.

Although prohibited from being installed as a permanent private sewage disposal system, cesspools may be necessary where permanent systems are under repair or are being built. Chapter 10 provides the details for constructing a cesspool.

Chapter 11 Residential Wastewater Systems.

Another method of private sewage disposal is a small wastewater treatment plant. Where permitted, these systems can discharge effluent directly to streams and rivers. Chapter 11 specifies the standard to which wastewater treatment plants must conform.

Chapter 12 Inspections.

The best soil and site analysis along with the best design will be rendered useless if the system is not installed according to the plans for the system. Chapter 12 provides requirements for inspection of private sewage disposal systems.

Chapter 13 Nonliquid Saturated Treatment Systems.

In some locations, water for the flushing of waste into and through a sanitary piping system is not available. For example, a toilet facility provided for a remote campground without running water would require such a system. Chapter 13 specifies the standard to which nonliquid saturated treatment systems must conform.

Chapter 14 Referenced Standards.

Chapter 14 lists all of the product and installation standards and codes that are referenced throughout Chapters 1 through 13 and includes identification of the promulgators and the section numbers in which the standards and codes are referenced. As stated in Section 102.10, these standards and codes become an enforceable part of the code (to the prescribed extent of the reference) as if printed in the body of the code.

Appendix A System Layout Illustrations.

Because each chapter of this code uses only words to describe requirements, illustrations can offer greater insight as to what the words mean. Appendix A has a number of illustrations referenced to specific sections of the code to help the reader gain a better understanding of the requirements of the code.

Appendix B Tables for Pressure Distribution Systems.

The design of a pressure distribution system is accomplished by the use of several complex formulas found in Chapter 7. Because a user of the code may not have the necessary experience to manipulate the formulas, a tabular approach for designing pressure distribution systems is provided in Appendix B.

Appendix C Board of Appeals.

Appendix C contains the provisions for appeal and the establishment of a board of appeals. The provisions include the application for an appeal, the makeup of the board of appeals and the conduct of the appeal process.

RELOCATION OF TEXT OR TABLES

The following table indicates relocation of sections and tables in the 2024 edition of the IPSDC from the 2021 edition.

RELOCATIONS—continued

2024 LOCATION	2021 LOCATION
104.2.2.4	105.3, 105.3.1
104.2.3	105.2
104.2.3.5	105.3.2
104.2.3.6	105.2.1
104.2.4	105.1
104.3	104.2
104.7.4	105.3.3
104.9	105.4
104.9.1	105.4.1
105	106
113	114
114	115

CONTENTS

CHAPTER 1

SCOPE AND ADMINISTRATION

User notes:

About this chapter: *Chapter 1 establishes the limits of applicability of this code and describes how the code is to be applied and enforced. Chapter 1 is in two parts: Part 1—Scope and Application (Sections 101–102) and Part 2—Administration and Enforcement (Sections 103–114). Section 101 identifies which buildings and structures come under its purview and references other I-Codes as applicable. Standards and codes are scoped to the extent referenced (see Section 102.10).*

This code is intended to be adopted as a legally enforceable document and it cannot be effective without adequate provisions for its administration and enforcement. The provisions of Chapter 1 establish the authority and duties of the code official appointed by the authority having jurisdiction and also establish the rights and privileges of the design professional, contractor and property owner.

Section 105 was revised and relocated to Section 104 for the 2024 edition. For clarity, the relocation marginal markings have not been included. For complete information, see the Relocations table in the Preface of this code.

QR code use: *A QR code is placed at the beginning of any section that has undergone technical revision. To see those revisions, scan the QR code with a smart device or enter the 7-digit code beneath the QR code at the end of the following URL: qr.iccsafe.org/ (see Formatting Changes to the 2024 International Codes for more information).*

PART 1—SCOPE AND APPLICATION

SECTION 101—SCOPE AND GENERAL REQUIREMENTS

[A] 101.1 Title. These regulations shall be known as the *Private Sewage Disposal Code* of [NAME OF JURISDICTION] hereinafter referred to as "this code."

[A] 101.2 Scope. Septic tank and effluent absorption systems or other treatment tank and effluent disposal systems shall be permitted where a public sewer is not available to the property served. Unless specifically approved, the *private sewage disposal system* of each building shall be entirely separate from and independent of any other building. The use of a common system or a system on a parcel other than the parcel where the structure is located shall be subject to the full requirements of this code as for systems serving public buildings.

[A] 101.2.1 Appendices. Provisions in the appendices shall not apply unless specifically adopted.

[A] 101.3 Public sewer connection. Where public sewers become available to the premises served, the use of the *private sewage disposal system* shall be discontinued within that period of time required by law, but such period shall not exceed 1 year. The building sewer shall be disconnected from the *private sewage disposal system* and connected to the public sewer.

[A] 101.4 Abandoned systems. Abandoned *private sewage disposal systems* shall be plugged or capped in an approved manner. Abandoned treatment tanks and *seepage pits* shall have the contents pumped and discarded in an approved manner. The top or entire tank shall be removed and the remaining portion of the tank or excavation shall be filled immediately.

[A] 101.5 Failing system. When a *private sewage disposal system* fails or malfunctions, the system shall be corrected or use of the system shall be discontinued within that period of time required by the *code official*, but such period shall not exceed 1 year.

[A] 101.5.1 Failure. A failing *private sewage disposal system* shall be one causing or resulting in any of the following conditions:

1. The failure to accept sewage discharge and backup of sewage into the structure served by the *private sewage disposal system*.
2. The discharge of sewage to the surface of the ground or to a drain tile.
3. The discharge of sewage to any surface or ground water.
4. The introduction of sewage into saturation zones adversely affecting the operation of a *private sewage disposal system*.

[A] 101.6 Purpose. The purpose of this code is to establish minimum requirements to provide a reasonable level of safety health, property protection and general welfare by regulating and controlling the design, construction, installation, quality of materials, location, operation and maintenance or use of *private sewage disposal systems*.

[A] 101.7 Severability. If any section, subsection, sentence, clause or phrase of this code is for any reason held to be unconstitutional, such decision shall not affect the validity of the remaining portions of this code.

SECTION 102—APPLICABILITY

[A] 102.1 General. Where there is a conflict between a general requirement and a specific requirement, the specific requirement shall govern. Where, in any specific case, different sections of this code specify different materials, methods of construction or other requirements, the most restrictive shall govern.

[A] 102.2 Other laws. The provisions of this code shall not be deemed to nullify any provisions of local, state or federal law.

[A] 102.3 Application of references. Reference to chapter section numbers, or to provisions not specifically identified by number, shall be construed to refer to such chapter, section or provision of this code.

[A] 102.4 Existing installations. *Private sewage disposal systems* lawfully in existence at the time of the adoption of this code shall be permitted to have their use and maintenance continued if the use, maintenance or repair is in accordance with the original design and no hazard to life, health or property is created by the system.

[A] 102.5 Maintenance. *Private sewage disposal systems*, materials and appurtenances, both existing and new, and all parts thereof shall be maintained in proper operating condition in accordance with the original design in a safe and sanitary condition. Devices or safeguards that are required by this code shall be maintained in compliance with the edition of the code under which they were installed. The owner or the owner's authorized agent shall be responsible for maintenance of *private sewage disposal systems*. To determine compliance with this provision, the *code official* shall have the authority to require reinspection of any *private sewage disposal system*.

[A] 102.6 Additions, alterations or repairs. Additions, alterations, renovations or repairs to any *private sewage disposal system* shall conform to that required for a new system without requiring the existing system to comply with all the requirements of this code. Additions, alterations or repairs shall not cause an existing system to become unsafe, insanitary or overloaded.

Minor additions, alterations, renovations and repairs to existing systems shall meet the provisions for new construction, unless such work is done in the same manner and arrangement as was in the existing system, is not hazardous and is approved.

[A] 102.7 Change in occupancy. It shall be unlawful to make any change in the occupancy of any structure that will subject the structure to any special provision of this code applicable to the new occupancy without approval of the *code official*. The *code official* shall certify that such structure meets the intent of the provisions of law governing building construction for the proposed new occupancy and that such change of occupancy does not result in any hazard to the public health, safety or welfare.

[A] 102.8 Historic buildings. The provisions of this code relating to the construction, alteration, repair, enlargement, restoration, relocation or moving of buildings or structures shall not be mandatory for existing buildings or structures identified and classified by the state or local jurisdiction as historic buildings when such buildings or structures are judged by the *code official* to be safe and in the public interest of health, safety and welfare regarding any proposed construction, alteration, repair, enlargement, restoration, relocation or moving of buildings.

[A] 102.9 Moved buildings. Except as determined by Section 102.4, *private sewage disposal systems* that are a part of buildings or structures moved into or within the jurisdiction shall comply with the provisions of this code for new installations.

[A] 102.10 Referenced codes and standards. The codes and standards referenced in this code shall be those that are listed in Chapter 14 and such codes and standards shall be considered to be part of the requirements of this code to the prescribed extent of each such reference and as further regulated in Sections 102.10.1 and 102.10.2.

Exception: Where enforcement of a code provision would violate the conditions of the listing of the equipment or appliance, the conditions of the listing and the manufacturer's installation instructions shall apply.

[A] 102.10.1 Conflicts. Where conflicts occur between provisions of this code and the referenced standards, the provisions of this code shall apply.

[A] 102.10.2 Provisions in referenced codes and standards. Where the extent of the reference to a referenced code or standard includes subject matter that is within the scope of this code, the provisions of this code, as applicable, shall take precedence over the provisions in the referenced code or standard.

[A] 102.11 Requirements not covered by code. Any requirements necessary for the proper operation of an existing or proposed *private sewage disposal system*, or for the public safety, health and general welfare, not specifically covered by this code, shall be determined by the *code official*.

PART 2—ADMINISTRATION AND ENFORCEMENT

SECTION 103—CODE COMPLIANCE AGENCY

[A] 103.1 Creation of agency. The **[INSERT NAME OF DEPARTMENT]** is hereby created and the official in charge thereof shall be known as the *code official*. The function of the agency shall be the implementation, administration and enforcement of the provisions of this code.

[A] 103.2 Appointment. The *code official* shall be appointed by the chief appointing authority of the jurisdiction.

[A] 103.3 Deputies. In accordance with the prescribed procedures of this jurisdiction and with the concurrence of the appointing authority, the *code official* shall have the authority to appoint a deputy *code official*, other related technical officers, inspectors and other employees. Such employees shall have powers as delegated by the *code official*.

SECTION 104—DUTIES AND POWERS OF THE CODE OFFICIAL

[A] 104.1 General. The *code official* is hereby authorized and directed to enforce the provisions of this code.

[A] 104.2 Determination of compliance. The code official shall have the authority to determine compliance with this code, to render interpretations of this code and to adopt policies and procedures in order to clarify the application of its provisions. Such interpretations, policies and procedures:

1. Shall be in compliance with the intent and purpose of this code.
2. Shall not have the effect of waiving requirements specifically provided for in this code.

[A] 104.2.1 Listed compliance. Where this code or a referenced standard requires equipment, materials, products or services to be listed and a listing standard is specified, the listing shall be based on the specified standard. Where a listing standard is not specified, the listing shall be based on an approved listing criteria. Listings shall be germane to the provision requiring the listing. Installation shall be in accordance with the listing and the manufacturer's instructions, and where required to verify compliance, the listing standard and manufacturer's instructions shall be made available to the code official.

[A] 104.2.2 Technical assistance. To determine compliance with this code, the code official is authorized to require the owner or owner's authorized agent to provide a technical opinion and report.

[A] 104.2.2.1 Costs. A technical opinion and report shall be provided without charge to the jurisdiction.

[A] 104.2.2.2 Preparer qualifications. The technical opinion and report shall be prepared by a qualified engineer, specialist, laboratory or specialty organization acceptable to the code official. The code official is authorized to require design submittals to be prepared by, and bear the stamp of, a registered design professional.

[A] 104.2.2.3 Content. The technical opinion and report shall analyze the properties of the design, operation or use of the building or premises and the facilities and appurtenances situated thereon to identify and propose necessary recommendations.

[A] 104.2.2.4 Tests. Where there is insufficient evidence of compliance with the provisions of this code, the code official shall have the authority to require tests as evidence of compliance. Test methods shall be as specified in this code or by other recognized test standards. In the absence of recognized test standards, the code official shall approve the testing procedures. Such tests shall be performed by a party acceptable to the code official.

[A] 104.2.3 Alternative materials, design and methods of construction and equipment. The provisions of this code are not intended to prevent the installation of any material or to prohibit any design or method of construction not specifically prescribed by this code, provided that any such alternative is not specifically prohibited by this code and has been approved.

Exception: Performance-based alternative materials, designs or methods of construction and equipment complying with the *International Code Council Performance Code*.

[A] 104.2.3.1 Approval authority. An alternative material, design or method of construction shall be approved where the code official finds that the proposed alternative is satisfactory and complies with Sections 104.2.3 through 104.2.3.7, as applicable.

[A] 104.2.3.2 Application and disposition. Where required, a request to use an alternative material, design or method of construction shall be submitted in writing to the code official for approval. Where the alternative material, design or method of construction is not approved, the code official shall respond in writing, stating the reasons the alternative was not approved.

[A] 104.2.3.3 Compliance with code intent. An alternative material, design or method of construction shall comply with the intent of the provisions of this code.

[A] 104.2.3.4 Equivalency criteria. An alternative material, design or method of construction shall, for the purpose intended, be not less than the equivalent of that prescribed in this code with respect to all of the following, as applicable:

1. Quality.
2. Strength.
3. Effectiveness.
4. Durability.
5. Safety, other than fire safety.
6. Fire safety.

104.2.3.5 Tests. Tests conducted to demonstrate equivalency in support of an alternative material, design or method of construction application shall be of a scale that is sufficient to predict performance of the end use configuration. Tests shall be performed by a party acceptable to the code official.

[A] 104.2.3.6 Reports. Supporting documentation, where necessary to assist in the approval of materials or assemblies not specifically provided for in this code, shall comply with Sections 104.2.3.6.1 and 104.2.3.6.2.

[A] 104.2.3.6.1 Evaluation reports. Evaluation reports shall be issued by an approved agency and use of the evaluation report shall require approval by the code official for the installation. The alternate material, design or method of construction and product evaluated shall be within the scope of the code official's recognition of the approved agency. Criteria used for the evaluation shall be identified within the report and, where required, provided to the code official.

[A] 104.2.3.6.2 Other reports. Reports not complying with Section 104.2.3.6.1 shall describe criteria, including but not limited to any referenced testing or analysis, used to determine compliance with code intent and justify code equivalence. The report shall be prepared by a qualified engineer, specialist, laboratory or specialty organization acceptable to the code official. The code official is authorized to require design submittals to be prepared by, and bear the stamp of, a registered design professional.

[A] 104.2.3.7 Peer review. The code official is authorized to require submittal of a peer review report in conjunction with a request to use an alternative material, design or method of construction, prepared by a peer reviewer that is approved by the code official.

[A] 104.2.4 Modifications. Where there are practical difficulties involved in carrying out the provisions of this code, the code official shall have the authority to grant modifications for individual cases, provided that the code official shall first find that one or more special individual reasons make the strict letter of this code impractical, and that the modification is in compliance with the intent and purpose of this code and that such modification does not lessen health, accessibility, life and fire safety or structural requirements. The details of the written request for and action granting modifications shall be recorded and entered in the files of the department of building safety.

[A] 104.2.4.1 Flood hazard areas. The code official shall not grant modifications to any provision required in flood hazard areas as established by Section 1612.3 of the *International Building Code*, unless a determination has been made that:

1. A showing of good and sufficient cause that the unique characteristics of the size, configuration or topography of the site render the elevation standards of Section 1612 of the *International Building Code* inappropriate.
2. A determination that failure to grant the variance would result in exceptional hardship by rendering the lot undevelopable.
3. A determination that the granting of a variance will not result in increased flood heights, additional threats to public safety or extraordinary public expense; cause fraud on or victimization of the public; or conflict with existing laws or ordinances.
4. A determination that the variance is the minimum necessary to afford relief, considering the flood hazard.
5. Submission to the applicant of written notice specifying the difference between the design flood elevation and the elevation to which the building is to be built, stating that the cost of flood insurance will be commensurate with the increased risk resulting from the reduced floor elevation, and stating that construction below the design flood elevation increases risks to life and property.

[A] 104.3 Applications and permits. The *code official* shall receive applications, review *construction documents,* issue permits, inspect the premises for which such permits have been issued and enforce compliance with the provisions of this code.

[A] 104.3.1 Determination of substantially improved or substantially damaged existing buildings and structures in flood hazard areas. For applications for reconstruction, rehabilitation, repair, alteration, addition or other improvement of existing buildings or structures located in flood hazard areas, the code official shall determine if the proposed work constitutes substantial improvement or repair of substantial damage. Where the code official determines that the proposed work constitutes substantial improvement or repair of substantial damage, and where required by this code, the code official shall require the building to meet the requirements of Section 1612 of the *International Building Code* or Section R322 of *the International Residential Code*, as applicable.

[A] 104.4 Right of entry. Where it is necessary to make an inspection to enforce the provisions of this code, or where the code official has reasonable cause to believe that there exists in a structure or on any premises a condition that is contrary to or in violation of this code that makes the structure or premises unsafe, dangerous or hazardous, the code official is authorized to enter the structure or premises at all reasonable times to inspect or to perform the duties imposed by this code. If such structure or premises is occupied, the code official shall present credentials to the occupant and request entry. If such structure or premises is unoccupied, the code official shall first make a reasonable effort to locate the owner, the owner's authorized agent or other person having charge or control of the structure or premises and request entry. If entry is refused, the code official shall have recourse to every remedy provided by law to secure entry.

[A] 104.4.1 Warrant. Where the code official has first obtained a proper inspection warrant or other remedy provided by law to secure entry, an owner, the owner's authorized agent, occupant or person having charge, care or control of the structure or premises shall not fail or neglect, after a proper request is made as herein provided, to permit entry therein by the code official for the purposes of inspection and examination pursuant to this code.

[A] 104.5 Identification. The *code official* shall carry proper identification when inspecting structures or premises in the performance of duties under this code.

[A] 104.6 Notices and orders. The code official shall issue necessary notices or orders to ensure compliance with this code. Notices of violations shall be in accordance with Section 113.

[A] 104.7 Official records. The code official shall keep official records as required by Sections 104.7.1 through 104.7.5. Such official records shall be retained for not less than 5 years or for as long as the building or structure to which such records relate remains in existence, unless otherwise provided by other regulations.

[A] 104.7.1 Approvals. A record of approvals shall be maintained by the code official and shall be available for public inspection during business hours in accordance with applicable laws.

[A] 104.7.2 Inspections. The code official shall have the authority to conduct inspections, or shall accept reports of inspection by approved agencies or individuals. Reports of such inspections shall be in writing and be certified by a responsible officer of such approved agency or by the responsible individual. The code official shall keep a record of each inspection made, including notices and orders issued, showing the findings and disposition of each.

[A] 104.7.3 Code alternatives and modifications. Application for alternative materials, design and methods of construction and equipment in accordance with Section 104.2.3; modifications in accordance with Section 104.2.4; and documentation of the final decision of the code official for either shall be in writing and shall be retained in the official records.

[A] 104.7.4 Tests. The code official shall keep a record of tests conducted to comply with Sections 104.2.2.4 and 104.2.3.5.

[A] 104.7.5 Fees. The code official shall keep a record of fees collected and refunded in accordance with Section 106.

[A] 104.8 Liability. The code official, member of the board of appeals or employee charged with the enforcement of this code, while acting for the jurisdiction in good faith and without malice in the discharge of the duties required by this code or other pertinent law or ordinance, shall not thereby be rendered personally liable, either civilly or criminally, and is hereby relieved from personal liability for any damage accruing to persons or property as a result of any act or by reason of any act or omission in the discharge of official duties.

[A] 104.8.1 Legal defense. Any suit or criminal complaint instituted against any officer or employee because of an act performed by that officer or employee in the lawful discharge of duties under the provisions of this code or other laws or ordinances implemented through the enforcement of this code shall be defended by legal representatives of the jurisdiction until the final termination of the proceedings. The code official or any subordinate shall not be liable for costs in any action, suit or proceeding that is instituted in pursuance of the provisions of this code.

[A] 104.9 Approved materials and equipment. Materials, equipment and devices approved by the code official shall be constructed and installed in accordance with such approval.

[A] 104.9.1 Materials and equipment reuse. Materials, equipment and devices shall not be reused unless such elements are in good working condition and approved.

SECTION 105—PERMITS

[A] 105.1 Where required. Work on a *private sewage disposal system* shall not commence until a permit for such work has been issued by the *code official*.

[A] 105.1.1 Annual permit. Instead of an individual construction permit for each alteration to an already approved system or equipment or appliance installation, the *code official* is authorized to issue an annual permit upon application therefor to any person, firm or corporation regularly employing one or more qualified tradespersons in the building, structure or on the premises owned or operated by the applicant for the permit.

[A] 105.1.2 Annual permit records. The person to whom an annual permit is issued shall keep a detailed record of alterations made under such annual permit. The *code official* shall have access to such records at all times or such records shall be filed with the *code official* as designated.

[A] 105.2 Application for permit. Each application for a permit, with the required fee, shall be filed with the *code official* on a form furnished for that purpose and shall contain a general description of the proposed work and its location. The application shall contain a description of the type of system, the system location, the occupancy of all parts of the structure and all portions of the site or lot not covered by the structure, and such additional information as is required by the *code official*. The maximum number of bedrooms for residential occupancies shall be indicated.

[A] 105.2.1 Preliminary inspection. Before a permit is issued, the *code official* is authorized to inspect and evaluate the systems, equipment, buildings, devices, premises and spaces or areas to be used.

[A] 105.2.2 Time limitation of application. An application for a permit for any proposed work shall be deemed to have been abandoned 180 days after the date of filing, unless such application has been pursued in good faith or a permit has been issued; except that the *code official* shall have the authority to grant one or more extensions of time for additional periods not exceeding 180 days each. The extension shall be requested in writing and justifiable cause demonstrated.

[A] 105.2.3 Previous approvals. This code shall not require changes in the *construction documents*, construction or designated occupancy of a structure for which a lawful permit has been heretofore issued or otherwise lawfully authorized, and the construction of which has been pursued in good faith within 180 days after the effective date of this code and has not been abandoned.

[A] 105.2.4 Soil data. Soil test reports shall be submitted indicating *soil boring* and percolation test data related to the undisturbed and finished grade elevations, vertical elevation reference point and horizontal reference point. Surface elevations shall be given for all *soil borings*. Soil reports shall bear the signature of a soil tester.

[A] 105.2.5 Site plan. A site plan shall be filed showing to scale the location of all septic tanks, holding tanks or other treatment tanks; building sewers; wells; water mains; water service; streams and lakes; *flood hazard areas*; dosing or pumping chambers; distribution boxes; effluent systems; dual disposal systems; replacement system areas; and the location of all buildings or structures. Separating distances and dimensions shall be shown, including any distance to adjoining property. A vertical elevation reference point and a horizontal reference point shall be indicated. For other than single-family dwellings, grade slope with contours shall be shown for the grade elevation of the entire area of the soil absorption system and the area on all sides for a distance of 25 feet (7620 mm).

[A] 105.3 Permit issuance. The application, *construction documents* and other data filed by an applicant for permit shall be reviewed by the *code official.* If the *code official* finds that the proposed work conforms to the requirements of this code and all laws and ordinances applicable thereto, and that the fees specified in Section 106.1 have been paid, a permit shall be issued to the applicant. A *private sewage disposal system* permit shall not be transferable.

[A] 105.3.1 Approved construction documents. When the *code official* issues the permit where *construction documents* are required, the *construction documents* shall be endorsed in writing and stamped "APPROVED." Such approved *construction documents* shall not be changed, modified or altered without authorization from the *code official.* Work shall be done in accordance with the approved *construction documents.*

The *code official* shall have the authority to issue a permit for the construction of a part of a *private sewage disposal system* before the *construction documents* for the whole system have been submitted or approved, provided adequate information and detailed statements have been filed complying with all pertinent requirements of this code. The holder of such permit shall proceed at his or her own risk without assurance that the permit for the entire system will be granted.

[A] 105.3.2 Validity. The issuance of a permit or approval of *construction documents* shall not be construed to be a permit for, or an approval of, any violation of any of the provisions of this code or of other ordinances of the jurisdiction. No permit presuming to give authority to violate or cancel the provisions of this code shall be valid.

The issuance of a permit based on *construction documents* and other data shall not prevent the *code official* from thereafter requiring the correction of errors in said *construction documents* and other data or from preventing building operations being carried on thereunder when in violation of this code or of other ordinances of the jurisdiction.

[A] 105.3.3 Expiration. Every permit issued by the *code official* under the provisions of this code shall expire by limitation and become null and void if the work authorized by such permit is not commenced within 180 days from the date of the permit, or if the work authorized by such permit is suspended or abandoned at any time after the work is commenced for a period of 180 days. Before such work can be recommenced, a new permit shall first be obtained and the fee therefor shall be one-half the amount required for a new permit for such work, provided that no changes have been or will be made in the original *construction documents* for such work, and provided further that such suspension or abandonment has not exceeded 1 year.

[A] 105.3.4 Extensions. Any permittee holding an unexpired permit shall have the right to apply for an extension of the time within which the permittee will commence work under that permit where work cannot be commenced within the time required by this section for good and satisfactory reasons. The *code official* shall extend the time for action by the permittee for a period not exceeding 180 days if there is reasonable cause. No permit shall be extended more than once. The fee for an extension shall be one-half the amount required for a new permit for such work.

[A] 105.3.5 Suspension or revocation of permit. The *code official* shall have the authority to suspend or revoke a permit issued under the provisions of this code wherever the permit is issued in error or on the basis of incorrect, inaccurate or incomplete information, or in violation of any ordinance or regulation or any of the provisions of this code.

[A] 105.3.6 Posting of permit. The permit or a copy shall be kept on the site of the work until the completion of the project.

SECTION 106—FEES

[A] 106.1 Payment of fees. A permit shall not be valid until the fees prescribed by law have been paid. An amendment to a permit shall not be released until the additional fee, if any, has been paid.

[A] 106.2 Schedule of permit fees. Where work requires a permit, a fee for each permit shall be paid as required, in accordance with the schedule as established by the applicable governing authority.

[A] 106.3 Permit valuations. The applicant for a *permit* shall provide an estimated value of the work for which the permit is being issued at time of application. Such estimated valuations shall include the total value of work, including materials and labor, for which the *permit* is being issued, such as electrical, gas, mechanical, plumbing equipment and permanent systems. Where, in the opinion of the *building official,* the valuation is underestimated, the *permit* shall be denied, unless the applicant can show detailed estimates acceptable to the *building official.* The building official shall have the authority to adjust the final valuation for permit fees.

[A] 106.4 Work commencing before permit issuance. Any person who commences any work on a *private sewage disposal system* before obtaining the necessary permits shall be subject to a fee established by the code official that shall be in addition to the required permit fees.

[A] 106.5 Related fees. The payment of the fee for the construction, alteration, removal or demolition for work done in connection to or concurrently with the work authorized by a permit shall not relieve the applicant or holder of the permit from the payment of other fees that are prescribed by law.

[A] 106.6 Refunds. The code official is authorized to establish a refund policy.

SECTION 107—CONSTRUCTION DOCUMENTS

[A] 107.1 Construction documents. An application for a permit shall be accompanied by not less than two copies of *construction documents* drawn to scale, or in a digital format where allowed by the building official, with sufficient clarity and detail dimensions showing the nature and character of the work to be performed. Specifications shall include pumps and controls, dose volume, elevation differences (vertical lift), pipe friction loss, pump performance curve, pump model and pump manufacturer. The *code official* is permitted to waive the requirements for filing *construction documents* where the work involved is of a minor nature. Where the quality of the materials is essential for conformity to this code, specific information shall be given to establish such quality, and this code shall not be cited, or the term "legal" or its equivalent used as a substitute for specific information.

[A] 107.2 Retention of construction documents. One set of approved *construction documents* shall be retained by the *code official* for a period of not less than 180 days from date of completion of the permitted work, or as required by state or local laws. One set of approved *construction documents* shall be returned to the applicant, and said set shall be kept on the site of the building or work at all times during which the work authorized thereby is in progress.

SECTION 108—NOTICE OF APPROVAL

[A] 108.1 Approval. After the prescribed inspections indicate that the work complies in all respects with this code, a notice of approval shall be issued by the *code official*.

[A] 108.1.1 Revocation. The *code official* is authorized to, in writing, suspend or revoke a notice of approval issued under the provisions of this code wherever the notice is issued in error, on the basis of incorrect information supplied, or where it is determined that the building or structure, premise or portion thereof is in violation of any ordinance or regulation or any of the provisions of this code.

SECTION 109—TEMPORARY USES, EQUIPMENT AND SYSTEMS

[A] 109.1 General. The *code official* is authorized to issue a permit for temporary uses, equipment, or systems. Such permits shall be limited as to time of service, but shall not be permitted for more than 180 days. The *code official* is authorized to grant extensions for demonstrated cause.

[A] 109.2 Conformance. Temporary uses, equipment and systems shall conform to the requirements of this code as necessary to ensure health, safety and general welfare.

[A] 109.3 Temporary utilities. The *code official* is authorized to give permission to temporarily supply service utilities in accordance with Section 110.

[A] 109.4 Termination of approval. The *code official* is authorized to terminate such permit for temporary uses, equipment or system and to order the same to be discontinued.

SECTION 110—SERVICE UTILITIES

[A] 110.1 Connection of service utilities. No person shall make connections from a utility, source of energy, fuel or power to any building or system that is regulated by this code for which a permit is required until authorized by the *code official*.

[A] 110.2 Temporary connection. The *code official* shall have the authority to authorize the temporary connection of the building or system to the utility, source of energy, fuel, water system or sewer system for the purpose of testing systems or for use under a temporary approval.

[A] 110.3 Authority to disconnect service utilities. The *code official* shall have the authority to authorize disconnection of utility service to the building, structure or system regulated by this code and the referenced codes and standards in case of emergency where necessary to eliminate an immediate hazard to life or property or where such utility connection has been made without the approval required by Section 110.1 or 110.2. The *code official* shall notify the serving utility, and wherever possible the owner or the owner's authorized agent and occupant of the building, structure or service system of the decision to disconnect prior to taking such action. If not notified prior to disconnecting, the owner, the owner's authorized agent or occupant of the building, structure or service system shall be notified in writing, as soon as practical thereafter.

SECTION 111—INSPECTIONS

[A] 111.1 Required inspections. After issuing a permit, the *code official* shall conduct inspections from time to time during and upon completion of the work for which a permit has been issued. A record of all such examinations and inspections and of all violations of this code shall be maintained by the *code official*.

[A] 111.1.1 Concealed work. It shall be the duty of the permit applicant to cause the work to remain visible and able to be accessed for inspection purposes. Neither the *code official* nor the jurisdiction shall be liable for expense entailed in the removal or replacement of any material required to allow inspection.

[A] 111.1.2 Other inspections. The *code official* is authorized to make or require other inspections to ascertain compliance with the provisions of this code and other laws that are enforced by the department.

[A] 111.1.3 Approved inspection agencies. The *code official* shall accept reports of approved inspection agencies provided that such agencies satisfy the requirements as to qualifications and reliability.

[A] 111.2 Special inspections. Special inspections of alternative engineered design *private sewage disposal systems* shall be conducted in accordance with Sections 111.2.1 and 111.2.2.

[A] 111.2.1 Periodic inspection. The *registered design professional* or designated inspector shall periodically inspect and observe the alternative engineered design to determine that the installation is in accordance with the approved plans. Discrepancies shall be brought to the immediate attention of the *private sewage disposal system* contractor for correction. Records shall be kept of all inspections.

[A] 111.2.2 Written report. The *registered design professional* shall submit a final report in writing to the *code official* upon completion of the installation, certifying that the alternative engineered design conforms to the approved *construction documents*. A notice of approval for the *private sewage disposal system* shall not be issued until a written certification has been submitted.

[A] 111.3 Contractor's responsibilities. It shall be the duty of every contractor who enters into contracts for the installation or repair of *private sewage disposal systems* for which a permit is required to comply with adopted state and local rules and regulations concerning licensing.

[A] 111.3.1 Inspection requests. It shall be the duty of the holder of the permit or their duly authorized agent to notify the *code official* when work is ready for inspection. It shall be the duty of the permit holder to provide access to and means for inspections of such work that is required by this code.

[A] 111.4 Approval required. Work shall not be done beyond the point indicated in each successive inspection without first obtaining the approval of the *code official*. The *code official*, upon notification, shall make the requested inspections and shall either indicate the portion of the construction that is satisfactory as completed, or notify the permit holder or his or her agent wherein the same fails to comply with this code. Any portions that do not comply shall be corrected and such portion shall not be covered or concealed until authorized by the *code official*.

[A] 111.5 Evaluation and follow-up inspection services. Prior to the approval of a prefabricated construction assembly having concealed work and the issuance of a permit, the *code official* shall require the submittal of an evaluation report on each prefabricated construction assembly, indicating the complete details of the *private sewage disposal system*, including a description of the system and its components, the basis on which the system is being evaluated, test results and similar information and other data as necessary for the *code official* to determine conformance to this code.

[A] 111.5.1 Evaluation service. The *code official* shall designate the evaluation service of an approved agency as the evaluation agency, and review such agency's evaluation report for adequacy and conformance to this code.

[A] 111.5.2 Follow-up inspection. Except where ready access is provided to *private sewage disposal systems*, service equipment and accessories for complete inspection at the site without disassembly or dismantling, the *code official* shall conduct the in-plant inspections as frequently as necessary to ensure conformance to the approved evaluation report or shall designate an independent, approved inspection agency to conduct such inspections. The inspection agency shall furnish the *code official* with the follow-up inspection manual and a report of inspections on request, and the installation shall have an identifying label permanently affixed to the system indicating that factory inspections have been performed.

[A] 111.5.3 Test and inspection records. Required test and inspection records shall be available to the *code official* at all times during the fabrication of the installation and the erection of the building; or such records as the *code official* designates shall be filed.

[A] 111.6 Testing. Installations shall be tested as required in this code and in accordance with Sections 111.6.1 through 111.6.3. Tests shall be made by the permit holder and observed by the *code official*.

[A] 111.6.1 New, altered, extended or repaired installations. New installations and parts of existing installations that have been altered, extended, renovated or repaired shall be tested as prescribed herein to disclose leaks and defects.

[A] 111.6.2 Apparatus, instruments, material and labor for tests. Apparatus, instruments, material and labor required for testing an installation or part thereof shall be furnished by the permit holder.

[A] 111.6.3 Reinspection and testing. Where any work or installation does not pass an initial test or inspection, the necessary corrections shall be made so as to achieve compliance with this code. The work or installation shall then be resubmitted to the *code official* for inspection and testing.

SECTION 112—MEANS OF APPEALS

[A] 112.1 General. In order to hear and decide appeals of orders, decisions or determinations made by the *code official* relative to the application and interpretation of this code, there shall be and is hereby created a board of appeals. The board of appeals shall be appointed by the applicable governing authority and shall hold office at its pleasure. The board shall adopt rules of procedure for conducting its business and shall render all decisions and findings in writing to the appellant with a duplicate copy to the *code official*.

112.2 Limitations on authority. An application for appeal shall be based on a claim that the true intent of this code or the rules legally adopted thereunder have been incorrectly interpreted, the provisions of this code do not fully apply or an equivalent or better form of construction is proposed. The board shall not have authority to waive requirements of this code.

[A] 112.3 Qualifications. The board of appeals shall consist of members who are qualified by experience and training on matters pertaining to the provisions of this code and are not employees of the jurisdiction.

[A] 112.4 Administration. The *code official* shall take action without delay in accordance with the decision of the board.

SECTION 113—VIOLATIONS

[A] 113.1 Unlawful acts. It shall be unlawful for any person, firm or corporation to erect, construct, alter, repair, remove, demolish or use any *private sewage disposal system*, or cause same to be done, in conflict with or in violation of any of the provisions of this code.

[A] 113.2 Notice of violation. The *code official* shall serve a notice of violation or order to the person responsible for the erection, installation, alteration, extension, repair, removal or demolition of private sewage disposal work in violation of the provisions of this code; in violation of a detailed statement or the approved *construction documents* thereunder or in violation of a permit or certificate issued under the provisions of this code. Such order shall direct the discontinuance of the illegal action or condition and the abatement of the violation.

[A] 113.3 Prosecution of violation. If the notice of violation is not complied with promptly, the *code official* shall request the legal counsel of the jurisdiction to institute the appropriate proceeding at law or in equity to restrain, correct or abate such violation, or to require the removal or termination of the unlawful system in violation of the provisions of this code or of the order or direction made pursuant thereto.

[A] 113.4 Violation penalties. Any person who shall violate a provision of this code or fail to comply with any of the requirements thereof or who shall erect, install, alter or repair private sewage disposal work in violation of the approved *construction documents* or directive of the *code official*, or of a permit or certificate issued under the provisions of this code, shall be guilty of a **[SPECIFY OFFENSE]**, punishable by a fine of not more than **[AMOUNT]** dollars or by imprisonment not exceeding **[NUMBER OF DAYS]**, or both such fine and imprisonment. Each day that a violation continues after due notice has been served shall be deemed a separate offense.

[A] 113.5 Abatement of violation. The imposition of the penalties herein prescribed shall not preclude the legal officer of the jurisdiction from instituting appropriate action to prevent unlawful construction or to restrain, correct or abate a violation; to prevent illegal occupancy of a building, structure or premises or to stop an illegal act, conduct, business or use of the *private sewage disposal system* on or about any premises.

[A] 113.6 Unsafe systems. Any *private sewage disposal system* regulated by this code that is unsafe or constitutes a health hazard, insanitary condition or is otherwise dangerous to human life is hereby declared unsafe. Any use of *private sewage disposal systems* regulated by this code constituting a hazard to safety, health or public welfare by reason of inadequate maintenance, dilapidation, obsolescence, disaster, damage or abandonment is hereby declared an unsafe use. Any such unsafe equipment is hereby declared to be a public *nuisance* and shall be abated by repair, rehabilitation, demolition or removal.

[A] 113.6.1 Authority to condemn equipment. Where the *code official* determines that any *private sewage disposal system,* or portion thereof, regulated by this code has become hazardous to life, health or property or has become insanitary, the *code official* shall order in writing that such system be either removed or restored to a safe or sanitary condition. A time limit for compliance with such order shall be specified in the written notice. A defective *private sewage disposal system* shall not be used or maintained after receiving such notice. Where such system is to be disconnected, written notice as prescribed in Section 113.2 shall be given. In cases of immediate danger to life or property, such disconnection shall be made immediately without such notice.

[A] 113.6.2 Authority to disconnect service utilities. The *code official* shall have the authority to authorize disconnection of utility service in accordance with Section 110.3.

SECTION 114—STOP WORK ORDER

[A] 114.1 Authority. Where the *code official* finds any work regulated by this code being performed in a manner contrary to the provisions of this code or in a dangerous or unsafe manner, the *code official* is authorized to issue a stop work order.

[A] 114.2 Issuance. The stop work order shall be in writing and shall be given to the owner of the property, the owner's authorized agent or the person performing the work. Upon issuance of a stop work order, the cited work shall immediately cease. The stop work order shall state the reason for the order and the conditions under which the cited work is authorized to resume.

[A] 114.3 Emergencies. Where an emergency exists, the *code official* shall not be required to give a written notice prior to stopping the work.

[A] 114.4 Failure to comply. Any person who shall continue any work after having been served with a stop work order, except such work as that person is directed to perform to remove a violation or unsafe condition, shall be subject to fines established by the authority having jurisdiction.

CHAPTER

2 DEFINITIONS

User notes:

About this chapter: *Codes, by their very nature, are technical documents. Every word, term and punctuation mark can add to or change the meaning of a technical requirement. It is necessary to maintain a consensus on the specific meaning of each term contained in the code. Chapter 2 performs this function by stating clearly what specific terms mean for the purpose of the code.*

SECTION 201—GENERAL

201.1 Scope. Unless otherwise expressly stated, the following words and terms shall, for the purposes of this code, have the meanings indicated in this chapter.

201.2 Interchangeability. Words used in the present tense include the future; words in the masculine gender include the feminine and neuter; the singular number includes the plural and the plural, the singular.

201.3 Terms defined in other codes. Where terms are not defined in this code and are defined in the *International Building Code* or the *International Plumbing Code*, such terms shall have meanings ascribed to them as in those codes.

201.4 Terms not defined. Where terms are not defined through the methods authorized by this section, such terms shall have ordinarily accepted meanings such as the context implies.

SECTION 202—GENERAL DEFINITIONS

AGGREGATE. Graded hard rock that has been washed with water under pressure over a screen during or after grading to remove fine material and with a hardness value of 3 or greater on Mohs' Scale of Hardness. Aggregate that will scratch a copper penny without leaving any residual rock material on the coin has a hardness value of 3 or greater on Mohs' Scale of Hardness.

[P] AIR BREAK (Drainage System). A piping arrangement in which a drain from a fixture, appliance or device discharges indirectly into another fixture, receptacle or interceptor at a point below the flood level rim and above the trap seal.

ALLUVIUM. Soil deposited by floodwaters.

APPROVED AGENCY. An established and recognized organization that is regularly engaged in conducting tests, furnishing inspection services or furnishing product evaluation or certification where such organization has been approved by the code official.

BEDROCK. The rock that underlies soil material or is located at the earth's surface. Bedrock is encountered when the weathered in-place consolidated material, larger than 0.08 inch (2 mm) in size, is more than 50 percent by volume.

CESSPOOL. A covered excavation in the ground receiving sewage or other organic wastes from a drainage system that is designed to retain the organic matter and solids, permitting the liquids to seep into the soil cavities.

CLEAR-WATER WASTES. Cooling water and condensate drainage from refrigeration compressors and air-conditioning equipment, water used for equipment chilling purposes, liquid having no impurities or where impurities have been reduced below a minimum concentration considered harmful, and cooled condensate from steam-heating systems or other equipment.

[A] CODE OFFICIAL. The officer or other designated authority charged with administration and enforcement of this code or a duly authorized representative.

COLLUVIUM. Soil transported under the influence of gravity.

COLOR. The moist color of the soil based on Munsell soil color charts.

[A] CONSTRUCTION DOCUMENTS. All the written, graphic and pictorial documents prepared or assembled for describing the design, location and physical characteristics of the elements of the project necessary for obtaining a building permit. The construction drawings shall be drawn to an appropriate scale.

CONVENTIONAL SOIL ABSORPTION SYSTEM. A system employing gravity flow from the septic or other treatment tank and applying effluent to the soil through the use of a *seepage trench*, bed or pit.

[BS] DESIGN FLOOD ELEVATION. The elevation of the "design flood," including wave height, relative to the datum specified on the community's legally designated flood hazard map. In areas designated as Zone AO, the design flood elevation shall be the elevation of the highest existing grade of the *building's* perimeter plus the depth number (in feet) specified on the flood hazard map. In areas designated as Zone AO where a depth number is not specified on the map, the depth number shall be taken as being equal to 2 feet (610 mm).

DETAILED SOIL MAP. A map prepared by or for a state or federal agency participating in the National Cooperative Soil Survey showing soil series, type and phases at a scale of not more than 2,000 feet to the inch (24 m/mm) and which includes related explanatory information.

DOSING SOIL ABSORPTION SYSTEM. A system employing a pump or automatic siphon to elevate or distribute effluent to the soil through the use of a *seepage trench* or bed.

EFFLUENT. Liquid discharged from a septic or other treatment tank.

[BS] FLOOD HAZARD AREA. The greater of the following two areas:

1. The area within a flood plain subject to a 1-percent or greater chance of flooding in any given year.
2. The area designated as a flood hazard area on a community's flood hazard map or as otherwise legally designated.

HIGH GROUND WATER. Soil saturation zones, including perched water tables, shallow regional ground water tables or aquifers, or zones seasonally, periodically or permanently saturated.

HOLDING TANK. An approved water-tight receptacle for collecting and holding sewage.

HORIZONTAL REFERENCE POINT. A stationary, easily identifiable point to which horizontal dimensions are related.

LEGAL DESCRIPTION. An accurate metes and bounds description, a lot and block number in a recorded subdivision, a recorded assessor's plat or a public land survey description to the nearest 40 acres (16 ha).

MANHOLE. An opening of sufficient size to permit a person to gain access to a sewer or any portion of a *private sewage disposal system*.

MOBILE UNIT. A structure of vehicular, portable design, built on a chassis and designed to be moved from one site to another and to be used with or without a permanent foundation.

MOBILE UNIT PARK. Any plot or plots of ground owned by a person, state or local government upon which two or more units, occupied for dwelling or sleeping purposes regardless of mobile unit ownership, are located and whether or not a charge is made for such accommodation.

[P] NUISANCE. Public *nuisance* as known in common law or equity jurisprudence; whatever is dangerous to human life or detrimental to health; whatever building, structure or premises is not sufficiently ventilated, sewered, drained, cleaned or lighted, in reference to its intended use; and whatever renders the air, human food, drink or water supply unwholesome.

PAN. A soil horizon cemented with any one of a number of cementing agents such as iron, organic matter, silica, calcium, carbonate, gypsum or a combination of chemicals. Pans will resist penetration from a knife blade and are slowly permeable horizons or are impermeable.

PEER REVIEW. An independent and objective technical review conducted by an approved third party.

PERCOLATION TEST. The method of testing absorption qualities of the soil (see Section 404).

PERMEABILITY. The ease with which liquids move through the soil. One of the soil qualities listed in soil survey reports.

PRESSURE DISTRIBUTION SYSTEM. A soil absorption system using a pump or automatic siphon and smaller diameter distribution piping with small-diameter perforations to introduce effluent into the soil.

PRIVATE SEWAGE DISPOSAL SYSTEM. A sewage treatment and disposal system serving a single structure with a septic tank and soil absorption field located on the same parcel as the structure. This term also means an alternative sewage disposal system, including a substitute for the septic tank or soil absorption field, a holding tank, a system serving more than one structure or a system located on a different parcel than the structure. A *private sewage disposal system* is permitted to be owned by the property owner or a special purpose district.

PRIVY. A structure, not connected to a plumbing system, that is used by persons for the deposition of human body waste.

[A] REGISTERED DESIGN PROFESSIONAL. An individual who is registered or licensed to practice their respective design profession, as defined by the statutory requirements of the professional registration laws of the state or jurisdiction in which the project is to be constructed.

SEEPAGE BED. An excavated area more than 5 feet (1524 mm) wide that contains a bedding of aggregate and has more than one distribution line.

SEEPAGE PIT. An underground receptacle constructed to permit disposal of effluent or clear wastes by soil absorption through its floor and walls.

SEEPAGE TRENCH. An area excavated 1 foot to 5 feet (305 mm to 1524 mm) wide containing a bedding of aggregate and a single distribution line.

SEPTAGE. All sludge, scum, liquid and any other material removed from a private sewage treatment and disposal system.

SEPTIC TANK. A tank that receives and partially treats sewage through processes of sedimentation, flotation and bacterial action to separate solids from the liquid in the sewage, and which discharges the liquid to a soil absorption system.

SOIL. The unconsolidated material over bedrock, 0.08 inch (2 mm) and smaller.

SOIL BORING. An observation pit dug by hand or backhoe, a hole dug by augering or a soil core taken intact and undisturbed with a probe.

SOIL MOTTLES. Spots, streaks or contrasting soil colors usually caused by soil saturation for one period of a normal year, with a color value of 4 or more and a chroma of 2 or less. Gray-colored mottles are called low chroma; reddish-brown, red- and yellow-colored mottles are called high chroma.

SOIL SATURATION. The state in which all pores in a soil are filled with water. Water will flow from saturated soil into a bore hole.

VENT CAP. An approved appurtenance used for covering the vent terminal of an effluent disposal system to avoid closure by mischief or debris and still permit circulation of air within the system.

VERTICAL ELEVATION REFERENCE POINT. An easily identifiable stationary point or object of constant elevation for establishing the relative elevation of percolation tests, *soil borings* and other locations.

WATERCOURSE. A stream usually flowing in a particular direction, though it need not flow continually and is sometimes dry. A watercourse flows in a definite channel, with a bed, sides or banks, and usually discharges itself into some other stream or body of water. It must be something more than mere surface drainage over the entire face of a tract of land, occasioned by unusual freshets or other extraordinary cause. It does not include surface water such as rain or melting snow flowing from higher to lower ground through hollows or ravines in land that is at other times destitute of water. Such hollows or ravines are not, in legal contemplation, watercourses.

WORKMANSHIP. Work of such character that will fully secure the results sought in all the sections of this code as intended for the health, safety and welfare protection of all individuals.

CHAPTER

3 GENERAL REGULATIONS

User notes:

About this chapter: *Chapter 3 covers general regulations for private sewage disposal installations. As many of these requirements would need to be repeated in Chapters 3 through 13, placing such requirements in only one location eliminates code development coordination issues associated with the same requirement in multiple locations. These general requirements can be superseded by more specific requirements for certain applications in Chapters 3 through 13.*

SECTION 301—GENERAL

301.1 Scope. The provisions of this chapter shall govern the general regulations of *private sewage disposal systems*, including specific limitations and *flood hazard areas*.

SECTION 302—SPECIFIC LIMITATIONS

302.1 Domestic waste. Waste and sewage derived from ordinary living uses shall enter the septic or treatment tank unless otherwise specifically exempted by the *code official* or this code.

302.2 Cesspools and privies. Privies shall be prohibited. *Cesspools* shall be prohibited, except where approved by the *code official*. Where approved, *cesspools* shall be designed and installed in accordance with Chapter 10.

302.3 Industrial wastes. The *code official* shall approve the method of treatment and disposal of all waste products from manufacturing or industrial operations, including combined industrial and domestic waste.

302.4 Detrimental or dangerous waste. Material such as ashes, cinders or rags; flammable, poisonous or explosive liquids or gases; oil, grease or other insoluble material that is capable of obstructing, damaging or overloading the *private sewage disposal system*, or is capable of interfering with the normal operation of the *private sewage disposal system*, shall not be deposited, by any means, into such systems. The *code official* shall approve the method of treatment and disposal.

302.5 Clear water. The discharge of surface, rain or other clear water into a *private sewage disposal system* shall be prohibited.

302.6 Water softener and iron filter backwash. Water softener or iron filter discharge shall be indirectly connected by means of an air gap to the *private sewage disposal system* or discharge onto the ground surface, provided that a *nuisance* is not created.

302.7 Food waste disposals. Where a food waste disposal connects to a *private sewage disposal system*, the system shall be designed to accommodate the solids loading from the disposal unit.

SECTION 303—FLOOD HAZARD AREAS

[BS] 303.1 General. Soil absorption systems shall be located outside of *flood hazard areas*.

> **Exception:** Where suitable soil absorption sites outside of the *flood hazard area* are not available, the soil absorption site is permitted to be located within the *flood hazard area*. The soil absorption site shall be located to minimize the effects of inundation under conditions of the design flood.

[BS] 303.2 Tanks. In *flood hazard areas*, tanks shall be anchored to counter buoyant forces during condition of the design flood. The vent termination and service manhole of the tank shall be not less than 2 feet (610 mm) above the *design flood elevation* or fitted with covers designed to prevent the inflow of floodwater or outflow of the contents of the tanks during conditions of the design flood.

[BS] 303.3 Mound systems. Mound systems shall be prohibited in *flood hazard areas*.

SECTION 304—ALTERNATIVE ENGINEERED DESIGN

304.1 Alternative engineered design. The design, documentation, inspection, testing and approval of an alternative engineered design *private sewage disposal system* shall comply with Sections 304.1.1 through 304.6.

304.1.1 Design criteria. An alternative engineered design shall conform to the intent of the provisions of this code and shall provide an equivalent level of quality, strength, effectiveness, fire resistance, durability and safety. Material, equipment or components shall be designed and installed in accordance with the manufacturer's instructions.

304.2 Submittal. The *registered design professional* shall indicate on the permit application that the *private sewage disposal system* is an alternative engineered design. The permit and permanent permit records shall indicate that an alternative engineered design was part of the approved installation.

304.3 Technical data. The *registered design professional* shall submit sufficient technical data to substantiate the proposed alternative engineered design and to prove that the performance meets the intent of this code.

304.4 Construction documents. The *registered design professional* shall submit to the *code official* two complete sets of signed and sealed *construction documents* for the alternative engineered design.

304.5 Design approval. Where the *code official* determines that the alternative engineered design conforms to the intent of this code, the *private sewage disposal system* shall be approved. If the alternative engineered design is not approved, the *code official* shall notify the *registered design professional* in writing, stating the reasons therefor.

304.6 Inspection and test. The alternative engineered design shall be inspected in accordance with the requirements of Section 111.

SITE EVALUATION AND REQUIREMENTS

User notes:

About this chapter: *Disposal systems covered in this code rely on the subsurface soil's abilities to accept the nonpotable water that is discharged by the treatment methods described in the code. Chapter 4 provides the methods for the evaluation of the soil in the planned disposal area.*

SECTION 401—GENERAL

401.1 Scope. The provisions of this chapter shall govern the evaluation of and requirements for *private sewage disposal system* sites.

401.2 Site evaluation. Site evaluation shall include soil conditions, properties and permeability, depth to zones of soil saturation, depth to bedrock, slope, landscape position, all setback requirements and the presence of *flood hazard areas*. Soil test data shall relate to the undisturbed elevations, and a vertical elevation reference point or benchmark shall be established. Evaluation data shall be reported on approved forms. Reports shall be filed within 30 days of the completion of testing for all sites investigated.

401.3 Replacement system area. On each parcel of land being initially developed, sufficient area of suitable soils based on the soil tests and system location and site requirements of this code for one replacement system shall be established. Where bore hole test data in the replacement system area are equivalent to data in the proposed system area, the percolation test is not required.

401.3.1 Nonconforming site conditions. Where site conditions do not permit replacement systems in accordance with this code and an alternative system is used, the alternative system shall be approved in accordance with Section 105.

401.3.2 Undisturbed site. The replacement system shall not be disturbed to the extent that the site area is no longer suitable. The replacement system area shall not be used for construction of buildings, parking lots or parking areas, below-ground swimming pools or any other use that will adversely affect the replacement area.

SECTION 402—SLOPE

402.1 General. A *conventional soil absorption system* shall not be located on land with a slope greater than 20 percent. A *conventional soil absorption system* shall be located not less than 20 feet (6096 mm) from the crown of land with a slope greater than 20 percent, except where the top of the aggregate of a system is at or below the bottom of an adjacent roadside ditch. Where a more restrictive land slope is to be observed for a soil absorption system, other than a *conventional soil absorption system*, the more restrictive land slope specified in the design sections of this code shall apply.

SECTION 403—SOIL BORINGS AND EVALUATION

403.1 Soil borings and profile descriptions. *Soil borings* shall be conducted on all sites, regardless of the type of private sewage system planned to serve the parcel. Borings shall extend not less than 3 feet (914 mm) below the bottom of the proposed system. Borings shall be of sufficient size and extent to determine the soil characteristics important to an on-site liquid waste disposal system. Borehole data shall be used to determine the suitability of soils at the site with respect to zones of seasonal or permanent soil saturation and the depth to bedrock. Borings shall be conducted prior to percolation tests to determine whether the soils are suitable to warrant such tests and, if suitable, at what depth percolation tests shall be conducted. The use of power augers for *soil borings* is prohibited. *Soil borings* shall be conducted and reported in accordance with Sections 403.1.1 through 403.1.5. Where it is not practical to have borings made with a backhoe, such borings shall be augered or dug by hand.

403.1.1 Number. There shall be not less than three borings per soil absorption site. Where necessary, more *soil borings* shall be made for an accurate evaluation of a site. Borings shall be constructed to a depth of not less than 3 feet (914 mm) below the proposed depth of the system.

Exception: On new parcels, the requirement of six borings (three for initial area and three for replacement area) shall be reduced to five where the initial and replacement system areas are contiguous and one boring is made on each outer corner of the contiguous area and the fifth boring is made between the system areas [see Appendix A, Figure A101.1(1)].

403.1.2 Location. Each borehole shall be accurately located and referenced to the vertical elevation and horizontal reference points. Reports of boring location shall either be drawn to scale or have the horizontal dimensions clearly indicated between the borings and the horizontal reference point.

403.1.3 Soil description. Soil profile descriptions shall be written for all borings. The thickness in inches (mm) of the different soil horizons observed shall be indicated. Horizons shall be differentiated on the basis of color, texture, *soil mottles* or bedrock. Depths shall be measured from the ground surface.

403.1.4 Soil mottles. Seasonal or periodic soil saturation zones shall be estimated at the highest level of *soil mottles*. The *code official* shall require, where deemed necessary, a detailed description of the soil mottling on a marginal site. The abundance, size, contrast and color of the *soil mottles* shall be described in the following manner:

Abundance shall be described as "few" if the mottled color occupies less than 2 percent of the exposed surface; "common" if the mottled color occupies from 2 to 20 percent of the exposed surface; or "many" if the mottled color occupies more than 20 percent of the exposed surface. Size refers to length of the mottle measured along the longest dimension and shall be described as "fine" if the mottle is less than 0.196 inch (5 mm); medium if the mottle is from 0.196 inch to 1.590 inches (5 mm to 40 mm); or coarse if the mottle is larger than 1.590 inches (40 mm). Contrast refers to the difference in color between the soil mottle and the background color of the soil and is described as "faint" if the mottle is evident but recognizable with close examination; "distinct" if the mottle is readily seen but not striking; or "prominent" if the mottle is obvious and one of the outstanding features of the horizon. The color(s) of the mottle(s) shall be indicated.

403.1.5 Observed ground water. The depth to ground water, if present, shall be reported. Observed ground water shall be reported at the level that ground water reaches in the soil borehole or the highest level of sidewall seepage into the boring. Measurements shall be made from ground level. Soil located above the water level in the boring shall be checked for the presence of *soil mottles*.

403.2 Color patterns not indicative of soil saturation. The following soil conditions shall be reported, but shall not be interpreted as color patterns caused by wetness or saturation. Soil profiles with an abrupt textural change with finer-textured soils overlying more than 4 feet (1219 mm) of unmottled, loamy sand or coarser soils can have a mottled zone for the finer textured material. Where the mottled zone is less than 12 inches (305 mm) thick and located immediately above the textural change, a soil absorption system shall be permitted in the loamy sand or coarser material below the mottled layer. The site shall be considered to be unsuitable where any *soil mottles* occur within the sandy material. The *code official* shall consider certain coarse sandy loam soils to be included as a coarse material.

403.2.1 Other soil color patterns. *Soil mottles* occur that are not caused by seasonal or periodic soil saturation zones. Examples of such soil conditions not limited by enumeration are *soil mottles* formed from residual sandstone deposits; *soil mottles* formed from uneven weathering of glacially deposited material or glacially deposited material that is naturally gray in color, including any concretionary material in various stages of decomposition; deposits of lime in a profile derived from highly calcareous parent material; light-colored silt coats deposited on soil bed faces; and *soil mottles* usually vertically oriented along old or decayed root channels with a dark organic stain usually present in the center of the mottled area.

403.2.2 Reporting exceptions. The site evaluator shall report any mottled soil condition. The observation of *soil mottles* not caused by soil saturation shall be reported. On request, the *code official* shall make a determination of the acceptability of the site.

403.3 Bedrock. The depth of the bedrock, except sandstone, shall be established at the depth in a *soil boring* where more than 50 percent of the weathered-in-place material is consolidated. Sandstone bedrock shall be established at the depth where an increase in resistance to penetration of a knife blade occurs.

403.4 Alluvial and colluvial deposits. Subsurface soil absorption systems shall not be placed in alluvial and colluvial deposits with shallow depths, extended periods of saturation or possible flooding.

SECTION 404—PERCOLATION OR PERMEABILITY EVALUATION

404.1 General. The permeability of the soil in the proposed absorption system shall be determined by percolation tests or permeability evaluation.

404.2 Percolation tests and procedures. Not less than three percolation tests in each system area shall be conducted. The holes shall be spaced uniformly in relation to the bottom depth of the proposed absorption system. More percolation tests shall be made where necessary, depending on system design.

404.2.1 Percolation test hole. The test hole shall be dug or bored. The test hole shall have vertical sides and a horizontal dimension of 4 inches to 8 inches (102 mm to 203 mm). The bottom and sides of the hole shall be scratched with a sharp-pointed instrument to expose the natural soil. Loose material shall be removed from the hole, and the bottom shall be covered with 2 inches (51 mm) of gravel or coarse sand.

404.2.2 Test procedure, sandy soils. The hole shall be filled with clear water to not less than 12 inches (305 mm) above the bottom of the hole for tests in sandy soils. The time for this amount of water to seep away shall be determined and this procedure shall be repeated if the water from the second filling of the hole seeps away in 10 minutes or less. The test shall proceed as follows: Water shall be added to a point not more than 6 inches (152 mm) above the gravel or coarse sand. Thereupon, from a fixed reference point, water levels shall be measured at 10-minute intervals for a period of 1 hour. Where 6 inches (152 mm) of water seeps away in less than 10 minutes, a shorter interval between measurements shall be used, but the water depth shall not exceed 6 inches (152 mm) in any case. Where 6 inches (152 mm) of water seeps away in less than 2 minutes, the test shall be stopped and a rate of less than 3 minutes per inch (7.2 s/mm) shall be reported. The final water level drop shall be used to calculate the percolation rate. Soils not meeting the requirements of this section shall be tested in accordance with Section 404.2.3.

404.2.3 Test procedure, other soils. The hole shall be filled with clear water, and a minimum water depth of 12 inches (305 mm) shall be maintained above the bottom of the hole for a 4-hour period by refilling whenever necessary or by use of an automatic siphon. Water remaining in the hole after 4 hours shall not be removed. Thereafter, the soil shall be allowed to swell not less than 16 hours or more than 30 hours. Immediately after the soil swelling period, the measurements for determining the percolation rate shall be made as follows: Any soil sloughed into the hole shall be removed, and the water level shall be adjusted to 6 inches (152 mm) above the gravel or coarse sand. Thereupon, from a fixed reference point, the water level shall be measured at 30-minute intervals for a period of 4 hours, unless two successive water level drops do not vary by more than $^{1}/_{16}$ inch (1.59 mm). Not less than three water level drops shall be observed and recorded. The hole shall be filled with clear water to a point not more than 6 inches (152 mm) above the gravel or coarse sand whenever it becomes nearly empty. The water level shall not be adjusted during the three measurement periods except to the limits of the last measured water level drop. Where the first 6 inches (152 mm) of water seeps away in less than 30 minutes, the test shall be performed again for a period of 1 hour with measurements performed every 10 minutes. The water depth shall not exceed 5 inches (127 mm) at any time during the measurement period. The drop that occurs during the final measurement period shall be used in calculating the percolation rate.

404.2.4 Mechanical test equipment. Mechanical percolation test equipment shall be of an approved type.

404.3 Permeability evaluation. Soil shall be evaluated for estimated percolation based on structure and texture in accordance with accepted soil evaluation practices. Borings shall be made in accordance with Section 404.2 for evaluating the soil.

SECTION 405—SOIL VERIFICATION

405.1 Verification. Where required by the *code official*, depth to *soil mottles*, depth to high ground water, soil textures, depth to bedrock and land slope shall be verified by the *code official*. The *code official* shall require, where necessary, backhoe pits to be provided for verification of *soil boring* data. Where required by the *code official*, the results of percolation tests or permeability evaluation shall be subject to verification. The *code official* shall require, where necessary, that percolation tests be conducted under supervision. Where the natural soil condition has been altered by filling or other methods used to improve wet areas, the *code official* shall require, where necessary, observation of high ground water levels under saturated soil conditions. Detailed soil maps, or other adequate information, shall be used for determining estimated percolation rates and other soil characteristics.

405.2 Monitoring ground water levels. A property owner or developer shall have the option to provide documentation that soil mottling or other color patterns at a particular site are not an indication of seasonally saturated soil conditions of high ground water levels. Direct observation shall be used to document ground water levels. Monitoring shall be in accordance with the procedures cited in Sections 405.2.1 through 405.2.6.

405.2.1 Precipitation. Monitoring shall be performed at a time of the year when maximum ground water elevation occurs. In determining whether a near-normal season has occurred where sites are subject to broad regional water tables, such as large areas of sandy soils, the fluctuation over the several-year cycle shall be considered. In such cases, data obtained from the United States Geological Survey (USGS) shall be used to determine if a regional water table was at or near its normal level.

405.2.2 Artificial drainage. Areas to be monitored shall be checked for drainage tile and open ditches that alter natural high ground water levels. Where such factors are involved, information on the location, design, ownership and maintenance responsibilities for such drainage shall be provided. Documentation shall be provided to show that the drainage network has an adequate outlet and will be maintained. Sites affected by agricultural drain tile shall not be acceptable for system installation.

405.2.3 Procedures. The owner or the owner's agent shall notify the *code official* in writing of the intent to monitor. Where necessary, the *code official* shall field check the monitoring once or more during the time of expected saturated soil conditions.

Not less than three wells shall be monitored at a site for a proposed system and replacement. Where necessary, the *code official* shall require more than three monitoring sites, and the site evaluator shall be so advised in writing.

405.2.4 Monitoring well design. Not less than two wells shall extend to a depth of not less than 6 feet (1829 mm) below the ground surface and shall be not less than 3 feet (914 mm) below the designed system depth. However, with layered mottled soil over permeable unmottled soil, not less than one well shall terminate within the mottled layer. Monitoring at greater depths shall be required, where necessary, due to site conditions. The site evaluator shall determine the depth of the monitoring wells for each specific site. Depths shall be approved. The monitoring well shall be a solid pipe installed in a bore hole. The pipe size shall be not less than 1 inch (25 mm) and not greater than 4 inches (102 mm). The bore hole shall be not less than 4 inches (102 mm) and not greater than 8 inches (203 mm) larger than the pipe (see Appendix A, Figure A101.1(2)).

405.2.5 Observations. The first observation shall be made on or before **[DATE]**. Observations shall be made thereafter every 7 days or less until **[DATE]** or until the site is determined to be unacceptable, whichever occurs first. Where water is observed above the critical depth at any time, an observation shall be made 1 week later. Where water is present above the critical depth at both observations, monitoring shall cease and the site shall be considered unacceptable. Where water is not present above the critical depth at the second observation, monitoring shall continue until **[DATE]**. Where any two observations 7 days apart show the presence of water above the critical depth, the site shall be considered unacceptable and the *code official* shall be notified in writing. When rainfall of 0.5 inch (12.7 mm) or more occurs in a 24-hour period during monitoring, observations shall be made at more frequent intervals, where necessary.

405.2.6 Reporting data. Where monitoring shows saturated conditions, the following data shall be submitted in writing: test locations; ground elevations at the wells; soil profile descriptions; soil series, if available from soil maps; dates observed; depths to observed water; and local precipitation data—monthly from **[DATE]** and daily during monitoring.

Where monitoring discloses that the site is acceptable, the following data shall be submitted in writing: location and depth of test holes, ground elevations at the wells and soil profile descriptions; soil series, if available from soil maps; dates observed; results of observations; information on artificial drainage; and local precipitation data—monthly from **[DATE]** and daily during monitoring. A request to install a soil absorption system shall be made in accordance with Section 105.

SECTION 406—SITE REQUIREMENTS

406.1 Soil absorption site location. The surface grade of all soil absorption systems shall be located at a point lower than the surface grade of any nearby water well or reservoir on the same or adjoining property. Where this is not possible, the site shall be located so surface water drainage from the site is not directed toward a well or reservoir. The soil absorption system shall be located with a minimum horizontal distance between various elements as indicated in Table 406.1. *Private sewage disposal systems* in compacted areas, such as parking lots and driveways, are prohibited. Surface water shall be diverted away from any soil absorption site on the same or neighboring lots.

TABLE 406.1—MINIMUM HORIZONTAL SEPARATION DISTANCES FOR SOIL ABSORPTION SYSTEMS

ELEMENT	DISTANCE (feet)
Cistern	50
Habitable building, below-grade foundation	25
Habitable building, slab-on-grade	15
Lake, high-water mark	50
Lot line	5
Reservoir	50
Roadway ditches	10
Spring	100
Streams or watercourse	50
Swimming pool	15
Uninhabited building	10
Water main	50
Water service	10
Water well	50

For SI: 1 foot = 304.8 mm.

406.1.1 Flood hazard areas. The site shall be located outside of *flood hazard areas*.

Exception: Where suitable sites outside of the *flood hazard area* are not available, it is permitted for the site to be located within the *flood hazard area*. The site shall be located to minimize the effects of inundation under conditions of the design flood.

406.2 Ground water, bedrock or slowly permeable soils. There shall be not less than 3 feet (914 mm) of soil between the bottom of the soil absorption system and high ground water or bedrock. Soil with a percolation rate of 60 minutes per 1 inch (25 mm) or faster shall exist for the depth of the proposed soil absorption system and not less than 3 feet (914 mm) below the proposed bottom of the soil absorption system. There shall be 56 inches (1422 mm) of suitable soil from original grade for a *conventional soil absorption system.*

406.3 Percolation rate, trench or bed. A subsurface soil absorption system of the trench or bed type shall not be installed where the percolation rate for any one of the three tests is slower than 60 minutes for water to fall 1 inch (25 mm). The slowest percolation rate shall be used to determine the absorption area.

406.4 Percolation rate, seepage pit. Percolation tests shall be made in each horizon penetrated below the inlet pipe for a *seepage pit*. Soil strata in which the percolation rates are slower than 30 minutes per 1 inch (25 mm) shall not be included in computing the absorption area. The slowest percolation rate shall be used to determine the absorption area.

406.5 Soil maps. Where a parcel of land consists entirely of soils with very severe or severe limitations for on-site liquidwaste disposal as determined by use of a detailed soil map and supporting data, that map and supporting data shall be permitted to be used as a basis for denial for an on-site waste disposal system. However, the property owner shall be permitted to present evidence that a suitable site for an on-site liquid-waste disposal system does exist.

406.6 Filled area. A soil absorption system shall not be installed in a filled area unless written approval is received.

406.6.1 Placement of fill. The approval of a *conventional soil absorption system* shall be based on evidence indicating its conformance to code requirements for area, percolation and elevation.

406.6.2 Bedrock. Where the original soil texture is sand or loamy sand, and the site has not less than 30 inches (762 mm) and not greater than 56 inches (1422 mm) of soil over bedrock, the fill shall be the same or coarser soil texture as the natural soil. Coarser fill material shall not be coarser than medium sand. Fill material shall not be finer than the natural soil.

406.6.3 High ground water. Sites with less than 56 inches (1422 mm) of soil over high ground water or estimated high ground water, where the original soil texture is sand or loamy sand, are permitted to be filled in accordance with Section 406.6.1 or 406.6.2.

406.6.4 Natural soil. Sites with soils finer than sand or loamy sand shall not be approved for systems in fill.

406.6.5 Monitoring. Sites that will have 36 inches (762 mm) or less of soil above high ground water after the top soil is removed shall be monitored for high ground water levels in the filled area in accordance with Section 405.2.

406.6.6 Inspection of fill. Placement of the fill material shall be inspected by the *code official*.

406.6.7 Design requirements. Filled areas shall be large enough to accommodate a shallow trench system and a replacement system. The site of the area to be filled shall be determined by the percolation rate of the natural soil and use of the building. Where any portion of the trench system or its replacement is in the fill, the fill shall extend 20 feet (6096 mm) beyond all sides of both systems before the slope begins. *Soil borings* and percolation tests shall be conducted before filling to determine soil textures and depth to high ground water or bedrock. Vegetation and topsoil shall be removed prior to filling. Slopes at the edge of the filled areas shall have a maximum ratio of one unit vertical to three units horizontal (33-percent slope), provided that the 20-foot (6096 mm) separating distance is maintained (see Appendix A, Figure A101.1(3)).

406.7 Altering slopes. Areas with slopes exceeding those specified in Section 402.1 shall not be used unless graded and reshaped in accordance with Sections 406.7.1 through 406.7.3.

406.7.1 Site investigation. Soil test data shall show that a sufficient depth of suitable soil material is present to provide the required amount of soil over bedrock and ground water after alteration. A complete site evaluation as specified in this section shall be performed after alteration of the site.

406.7.2 System location. A soil absorption system shall be installed in the cut area of an altered site. A soil absorption system shall not be installed in the fill area of an altered site. The area of fill on an altered site is permitted to be used as a portion of the required 20-foot (6096 mm) separating distance from the crown of a critical slope. There shall be not less than 6 feet (1829 mm) of natural soil between the edge of a system area and the downslope side of the altered area.

406.7.3 Site protection. Altered slope areas shall be positioned so that surface water drainage will be diverted away from the system areas. Disturbed areas shall be seeded or sodded with grass, and appropriate steps shall be taken to control erosion (see Figure 406.7.3).

FIGURE 406.7.3—CONCEPTUAL DESIGN SKETCH FOR ALTERING SLOPES

A. EXCAVATION OF COMPLETE HILLTOP

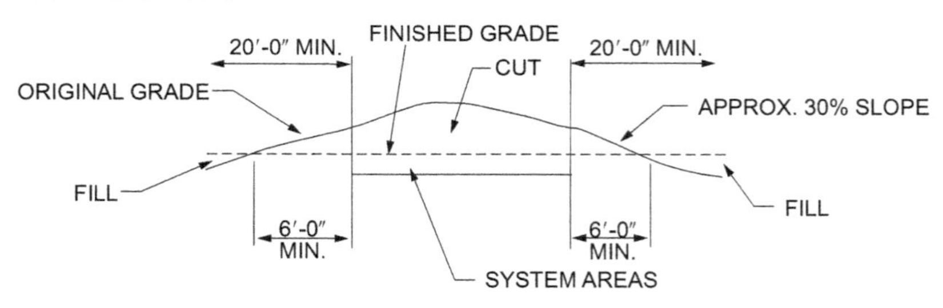

B. EXCAVATION INTO HILLSIDE

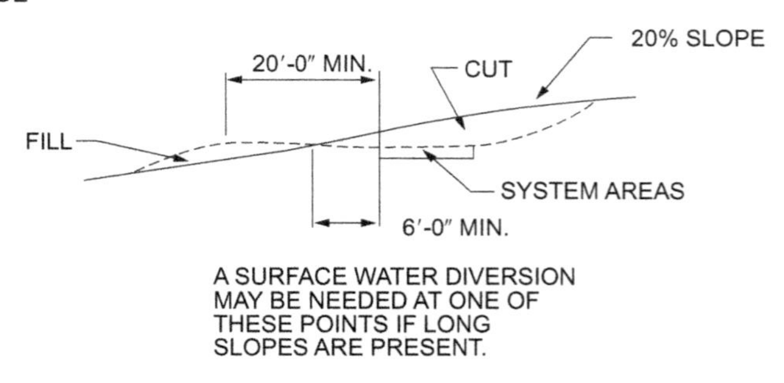

C. REGRADE OF HILLSIDE

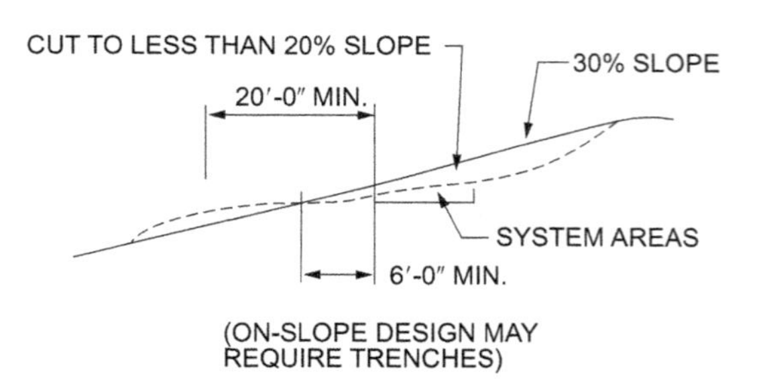

For SI: 1 inch = 25.4 mm, 1 foot = 304.8 mm.

CHAPTER 5 MATERIALS

User notes:

About this chapter: *Piping materials used in private sewage disposal systems must comply with standards. Chapter 5 indicates the standards for these products and specifies the material requirements for steel, concrete and fiberglass tanks.*

SECTION 501—GENERAL

501.1 Scope. The provisions of this chapter shall govern the requirements for materials for *private sewage disposal systems.*

501.2 Minimum standards. Materials shall conform to the standards referenced in this code for the construction, installation, alteration or repair of *private sewage disposal systems* or parts thereof.

> **Exception:** The extension, addition to or relocation of existing pipes with materials of like grade or quality in accordance with Sections 102.6 and 105.

SECTION 502—IDENTIFICATION

502.1 General. The manufacturer's mark or name and the quality of the product or identification shall be cast, embossed, stamped or indelibly marked on each length of pipe and each pipe fitting, fixture, tank, material and device used in a *private sewage disposal system* in accordance with the approved standard. Tanks shall indicate their capacity.

SECTION 503—PERFORMANCE REQUIREMENTS

503.1 Approved materials required. Materials, fixtures or equipment used in the installation, repair or alteration of any *private sewage disposal system* shall conform to the standards referenced in this code, except as otherwise approved in accordance with Section 105.

503.2 Care in installation. Materials installed in *private sewage disposal systems* shall be handled and installed so as to avoid damage. The quality of the material shall not be impaired.

503.3 Defective materials prohibited. Defective or damaged materials, equipment or apparatus shall not be installed or maintained.

SECTION 504—TANKS

504.1 Approval. Tanks shall be of an approved type. The design of tanks shall conform to the requirements of Chapter 8. Tanks shall be designed to withstand the pressures to which they are subjected.

504.1.1 Precast concrete and site-constructed tanks. Precast concrete septic tanks and square or rectangular holding tanks shall conform to ASTM C1277. The floor and sidewalls of a site-constructed concrete tank shall be monolithic, except a construction joint is permitted in the lower 12 inches (305 mm) of the sidewalls of the tank. The construction joint shall have a keyway in the lower section of the joint. The width of the keyway shall be approximately 30 percent of the thickness of the sidewall with a depth equal to the width. A continuous water stop or baffle not less than 56 inches (1422 mm) wide shall be set vertically in the joint, embedded one-half its width in the concrete below the joint with the remaining width in the concrete above the joint. The water stop or baffle shall be copper, neoprene, rubber or polyvinyl chloride designed for this specific purpose. Joints between the concrete septic tank and the tank cover and between the septic tank cover and manhole riser shall be tongue and groove or shiplap-type and sealed watertight using cement, mortar or bituminous compound. Connections between concrete septic tanks and holding tanks shall conform to ASTM C1644

504.1.1.1 Manhole covers. Manhole covers shall be of an approved material that maintains a watertight seal. When required by the jurisdiction having authority each manhole cover shall have an effective locking device.

504.1.1.2 Precast circular concrete. Precast circular concrete manhole riser sections, collars circular dosing or pump chambers, and holding tanks shall conform to ASTM C478.

504.1.1.3 Precast square or rectangular concrete. Precast square or rectangular concrete riser sections, collars, dosing or pump chambers shall conform to ASTM C913.

504.1.2 Steel tanks. Steel tanks shall conform to UL 70. Any damage to the bituminous coating shall be repaired by recoating. The gage of the steel shall be in accordance with Table 504.1.2.

TABLE 504.1.2 TANK CAPACITY

TANK DESIGN AND CAPACITY		MINIMUM GAGE THICKNESS	MINIMUM DIAMETER
Vertical cylindrical			
500 to 1,000 gallons	Bottom and sidewalls	12 gage	None
	Cover	12 gage	
	Baffles	12 gage	
1,001 to 1,250 gallons	Complete tank	10 gage	None
1,251 to 1,500 gallons	Complete tank	7 gage	None
Horizontal cylindrical			
500 to 1,000 gallons	Complete tank	12 gage	54-inch diameter
1,001 to 1,500 gallons	Complete tank	12 gage	64-inch diameter
1,501 to 2,500 gallons	Complete tank	10 gage	76-inch diameter
2,501 to 9,000 gallons	Complete tank	7 gage	76-inch diameter
9,001 to 12,000 gallons	Complete tank	$^{1}/_{4}$-inch plate	None
Over 12,000 gallons	Complete tank	$^{5}/_{16}$ inch	None

For SI: 1 inch = 25.4 mm, 1 gallon = 3.785 L.

504.1.3 Fiberglass tanks. Fiberglass tanks shall conform to ASTM D4021.

504.2 Manholes. Manhole collars and extensions shall be of an approved material that maintains a watertight seal.

504.2.1 Manhole covers. Manhole covers shall be of an approved material that maintains a watertight seal. Where required by the jurisdiction having authority, each manhole cover shall have an effective locking device.

SECTION 505—PIPE, JOINTS AND CONNECTIONS

505.1 Pipe. Pipe for *private sewage disposal systems* shall have a smooth wall and conform to one of the standards listed in Table 505.1.

TABLE 505.1—PRIVATE SEWAGE DISPOSAL SYSTEM PIPE

MATERIAL	STANDARD
Acrylonitrile butadiene styrene (ABS) plastic pipe	ASTM D2661; ASTM D2751; ASTM F628
Asbestos-cement pipe	ASTM C428
Cast-iron pipe	ASTM A74; ASTM A888; CISPI 301
Coextruded composite ABS DWV Schedule 40 IPS pipe (solid)	ASTM F1488; ASTM F1499
Coextruded composite ABS DWV Schedule 40 IPS pipe (cellular core)	ASTM F 1488; ASTM F1499
Coextruded composite ABS sewer and drain DR-PS in PS35, PS50, PS100, PS140 and PS200	ASTM F1488; ASTM F1499
Coextruded composite PVC DWV Schedule 40 IPS pipe (solid)	ASTM F1488
Coextruded composite PVC DWV Schedule 40 IPS pipe (cellular core)	ASTM F1488
Coextruded composite PVC-IPS-DR of PS140, PS200, DWV	ASTM F1488
Coextruded composite PVC 3.25OD DWV pipe	ASTM F1488
Coextruded composite PVC sewer and drain DR-PS in PS35, PS50, PS100, PS140 and PS200	ASTM F1488
Concrete pipe	ASTM C14; ASTM C76; CSA A257.1M; CSA A257.2M
Copper or copper-alloy tubing (Type K or L)	ASTM B75; ASTM B88; ASTM B251
Polyvinyl chloride (PVC) plastic pipe (Type DWV, SDR26, SDR35, SDR41, PS50 or PS 100)	ASTM D2665; ASTM D2949; ASTM D3034; ASTM F891; CSA B182.2; CSA B182.4
Vitrified clay pipe	ASTM C4; ASTM C700

505.1.1 Distribution pipe. Perforated pipe for distribution systems shall conform to one of the standards listed in Table 505.1 or Table 505.1.1.

TABLE 505.1.1—DISTRIBUTION PIPE

MATERIAL	STANDARD
Polyethylene (PE) plastic pipe	ASTM F405
Polyvinyl chloride (PVC) plastic pipe	ASTM D2729
Polyvinyl chloride (PVC) plastic pipe with pipe stiffness of PS35 and PS50	ASTM F1488

505.2 Joints and connection approval. Joints and connections shall be of an approved type.

505.3 ABS plastic pipe. Joints between acrylonitrile butadiene styrene (ABS) plastic pipe or fittings shall be in accordance with Sections 505.3.1 and 505.3.2.

505.3.1 Mechanical joints. Mechanical joints on drainage pipes shall be made with an elastomeric seal conforming to ASTM C1173, ASTM D3212 or CSA B602. Mechanical joints shall be installed only in underground systems, except as otherwise approved. Joints shall be installed in accordance with the manufacturer's instructions.

505.3.2 Solvent cementing. Joint surfaces shall be clean and free from moisture. Solvent cement conforming to ASTM D2235 or CSA B181.1 shall be applied to all joint surfaces. The joint shall be made while the cement is wet. Joints shall be made in accordance with ASTM D2235, ASTM D2661, ASTM F628 or CSA B181.1. Solvent cement joints shall be permitted above or below ground.

505.4 Asbestos-cement pipe. Joints between asbestos-cement pipe or fittings shall be made with a sleeve coupling of the same composition as the pipe and sealed with an elastomeric ring conforming to ASTM D1869.

505.5 Coextruded composite ABS pipe and joints. Joints between coextruded composite pipe with an ABS outer layer or ABS fittings shall comply with Sections 505.5.1 and 505.5.2.

505.5.1 Mechanical joints. Mechanical joints on drainage pipe shall be made with an elastomeric seal conforming to ASTM C1173, ASTM D3212 or CSA B602. Mechanical joints shall not be installed in above-ground systems, except as otherwise approved. Joints shall be installed in accordance with the manufacturer's instructions.

505.5.2 Solvent cementing. Joint surfaces shall be clean and free from moisture. Solvent cement conforming to ASTM D2235 or CSA B181.1 shall be applied to all joint surfaces. The joint shall be made while the cement is wet. Joints shall be made in accordance with ASTM D2235, ASTM D2661, ASTM F628 or CSA B181.1. Solvent cement joints shall be permitted above or below ground.

505.6 Cast-iron pipe. Joints between cast-iron pipe or fittings shall be in accordance with Sections 505.6.1 through 505.6.3.

505.6.1 Caulked joints. Joints for hub and spigot pipe shall be firmly packed with oakum or hemp. Molten lead shall be poured in one operation to a depth of not less than 1 inch (25 mm). The lead shall not recede more than 0.125 inch (3.2 mm) below the rim of the hub, and shall be caulked tight. Paint, varnish or other coatings shall not be applied to the joining material until after the joint has been tested and approved. Lead shall be run in one pouring and shall be caulked tight. Acid-resistant rope and acidproof cement shall be permitted.

505.6.2 Mechanical compression joints. Compression gaskets for hub and spigot pipe and fittings shall conform to ASTM C564. Gaskets shall be compressed when the pipe is fully inserted.

505.6.3 Mechanical joint coupling. Mechanical joint couplings for hubless pipe and fittings shall comply with CISPI 310 or ASTM C1277. The elastomeric sealing sleeve shall conform to ASTM C564 or CSA B602 and shall be provided with a center stop. Mechanical joint couplings shall be installed in accordance with the manufacturer's instructions.

505.7 Concrete pipe. Joints between concrete pipe or fittings shall be made by the use of an elastomeric seal conforming to ASTM C443, ASTM C1173, CSA A257.3M or CSA B602.

505.8 Copper or copper-alloy tubing or pipe. Joints between copper or copper-alloy tubing, pipe or fittings shall be in accordance with Sections 505.8.1 and 505.8.2.

505.8.1 Mechanical joints. Mechanical joints shall be installed in accordance with the manufacturer's instructions.

505.8.2 Soldered joints. Solder joints shall be made in accordance with the methods of ASTM B828. Cut ends shall be reamed to the full inside diameter of the tube end. All joint surfaces shall be cleaned. A flux conforming to ASTM B813 shall be applied. The joint shall be soldered with a solder conforming to ASTM B32.

505.9 Polyethylene plastic pipe and tubing. Joints between polyethylene plastic pipe and tubing or fittings shall be in accordance with Sections 505.9.1 and 505.9.2.

505.9.1 Heat-fusion joints. Joint surfaces shall be clean and free from moisture. Joint surfaces shall be heated to melting temperature and joined. The joint shall be undisturbed until cool. Joints shall be made in accordance with ASTM D2657.

505.9.2 Mechanical joints. Mechanical joints shall be installed in accordance with the manufacturer's instructions.

505.10 PVC plastic pipe. Joints between polyvinyl chloride (PVC) plastic pipe and fittings shall be in accordance with Sections 505.10.1 and 505.10.2.

505.10.1 Mechanical joints. Mechanical joints shall be made with an elastomeric seal conforming to ASTM C1173, ASTM D3212 or CSA B602. Mechanical joints shall not be installed in above-ground systems, except as otherwise approved. Joints shall be installed in accordance with the manufacturer's instructions.

505.10.2 Solvent cementing. Joint surfaces shall be clean and free from moisture. A purple primer that conforms to ASTM F656 shall be applied. Solvent cement not purple in color and conforming to ASTM D2564, CSA B137.3, CSA B181.2 or CSA B182.1 shall be applied to all joint surfaces. The joint shall be made while the cement is wet, and shall be in accordance with ASTM D2855. Solvent cement joints shall be permitted above or below ground.

505.11 Coextruded composite PVC pipe. Joints between coextruded composite pipe with a PVC outer layer or PVC fittings shall comply with Sections 505.11.1 and 505.11.2.

505.11.1 Mechanical joints. Mechanical joints on drainage pipe shall be made with an elastomeric seal conforming to ASTM D3212. Mechanical joints shall not be installed in above-ground systems, except as otherwise approved. Joints shall be installed in accordance with the manufacturer's instructions.

505.11.2 Solvent cementing. Joint surfaces shall be clean and free from moisture. A purple primer that conforms to ASTM F656 shall be applied. Solvent cement not purple in color and conforming to ASTM D2564, CSA B137.3, CSA B181.2 or CSA B182.1 shall be applied to all joint surfaces. The joint shall be made while the cement is wet, and shall be in accordance with ASTM D2855. Solvent cement joints shall be permitted above or below ground.

505.12 Vitrified clay pipe. Joints between vitrified clay pipe or fittings shall be made by the use of an elastomeric seal conforming to ASTM C425, ASTM C1173 or CSA B602.

505.13 Different piping materials. Joints between different piping materials shall be made with a mechanical joint of the compression or mechanical-sealing type conforming to ASTM C1173, ASTM C1460 or ASTM C1461. Connectors or adapters shall be approved for the application and such joints shall have an elastomeric seal conforming to ASTM C425, ASTM C443, ASTM C564, ASTM C1440, ASTM D1869, ASTM F477, CSA A257.3M or CSA B602 or as required in Sections 505.13.1 and 505.13.2. Joints shall be installed in accordance with the manufacturer's instructions.

505.13.1 Copper to cast-iron hub pipe. Joints between copper pipe or copper alloy tubing and cast-iron hub pipe shall be made with a copper-alloy ferrule or compression joint. The copper pipe or tubing shall be soldered to the ferrule in an approved manner, and the ferrule shall be joined to the cast-iron hub by a caulked joint or a mechanical compression joint.

505.13.2 Plastic pipe or tubing to other piping material. Joints between different grades of plastic pipe or between plastic pipe and other piping material shall be made with an approved adapter fitting. Joints between plastic pipe and cast-iron hub pipe shall be made by a caulked joint or a mechanical compression joint.

505.14 Pipe installation. Pipe shall be installed in accordance with the *International Plumbing Code*.

SECTION 506—PROHIBITED JOINTS AND CONNECTIONS

506.1 General. The following types of joints and connections shall be prohibited:

1. Cement or concrete joints.
2. Mastic or hot-pour bituminous joints.
3. Joints made with fittings not approved for the specific installation.
4. Joints between different diameter pipes made with elastomeric rolling O-rings.
5. Solvent-cement joints between different types of plastic pipe.

CHAPTER

6

SOIL ABSORPTION SYSTEMS

User notes:

About this chapter: *Chapter 6 covers the design of 'conventional' soil absorption systems. These systems are conventional in the sense that nonpotable water from the outlet of a septic tank flows by gravity into a piping network for distributing the water in an excavated area nearby. The piping is backfilled with the finished grade blending into adjacent grade level.*

SECTION 601—GENERAL

601.1 Scope. The provisions of this chapter shall govern the sizing and installation of soil absorption systems.

SECTION 602—SIZING SOIL ABSORPTION SYSTEMS

602.1 General. Effluent from septic tanks and other approved treatment tanks shall be disposed of by soil absorption or an approved manner. Sizing shall be in accordance with this chapter for systems with a daily effluent application of 5,000 gallons (18 925 L) or less. Two systems of equal size shall be required for systems receiving effluents exceeding 5,000 gallons (18 925 L) per day. Each system shall have a minimum capacity of 75 percent of the area required for a single system. An approved means of alternating waste application shall be provided. A dual system shall be considered as one system.

602.2 Pressure system. A *pressure distribution system* shall be permitted in place of a conventional or dosing *conventional soil absorption system* where a site is suitable for a conventional *private sewage disposal system*. A *pressure distribution system* shall be approved as an alternative *private sewage disposal system* where the site is unsuitable for conventional treatment (for sizing and design criteria, see Chapter 7).

602.3 Method of discharge. Flow from the septic or treatment tank to the soil absorption system shall be by gravity or dosing for facilities with a daily effluent application of 1,500 gallons (5678 L) or less. The tank effluent shall be discharged by pumping or an automatic siphon for systems over 1,500 gallons (5678 L).

SECTION 603—RESIDENTIAL SIZING

603.1 General. The bottom area for *seepage trenches* or beds or the sidewall area for *seepage pits* required for a soil absorption system serving residential property shall be determined from Table 603.1 using soil percolation test data and type of construction.

TABLE 603.1—MINIMUM ABSORPTION AREA FOR ONE- AND TWO-FAMILY DWELLINGS

PERCOLATION CLASS	PERCOLATION RATE (minutes required for water to fall 1 inch)	SEEPAGE TRENCHES OR PITS (square feet per bedroom)	SEEPAGE BEDS (square feet per bedroom)
1	0 to less than 10	165	205
2	10 to less than 30	250	315
3	30 to less than 45	300	375
4	45 to 60	330	415

For SI: 1 minute per inch = 2.4 s/mm, 1 square foot = 0.0929 m^2.

SECTION 604—OTHER BUILDING SIZING

604.1 General. The minimum required soil absorption system area for all occupancies, except one- and two-family dwellings, shall be based on building usage, the percolation rate and system design in accordance with Tables 604.1(1) and 604.1(2). The minimum soil absorption area shall be calculated by the following equation:

Equation 6-1 $A = U \times CF \times AA$

where:

A = Minimum system absorption area.

AA = Absorption area from Table 604.1(1).

CF = Conversion factor from Table 604.1(2).

U = Number of units.

TABLE 604.1(1)—MINIMUM ABSORPTION AREA FOR OTHER THAN ONE- AND TWO-FAMILY DWELLINGS			
PERCOLATION CLASS	**PERCOLATION RATE (minutes required for water to fall 1 inch)**	**SEEPAGE TRENCHES OR PITS (square feet per unit)**	**SEEPAGE BEDS (square feet per unit)**
1	0 to less than 10	110	140
2	10 to less than 30	165	205
3	30 to less than 45	220	250
4	45 to 60	220	280

For SI: 1 minute per inch = 2.4 s/mm, 1 square foot = 0.0929 m^2.

TABLE 604.1(2)—CONVERSION FACTOR		
BUILDING CLASSIFICATION	**UNITS**	**FACTOR**
Apartment building	1 per bedroom	1.5
Assembly hall—no kitchen	1 per person	0.02
Auto washer (service buildings, etc.)	1 per machine	6.0
Bar and cocktail lounge	1 per patron space	0.2
Beauty salon	1 per station	2.4
Bowling center	1 per bowling lane	2.5
Bowling center with bar	1 per bowling lane	4.5
Camp, day and night	1 per person	0.45
Camp, day use only	1 per person	0.2
Campground and camping resort	1 per camping space	0.9
Campground and sanitary dump station	1 per camping space	0.085
Car wash	1 per car	1.0
Catch basin—garages, motor-fuel-dispensing facility, etc.	1 per basin	2.0
Catch basin—truck wash	1 per truck	5.0
Church—no kitchen	1 per person	0.04
Church—with kitchen	1 per person	0.09
Condominium	1 per bedroom	1.5
Dance hall	1 per person	0.06
Dining hall—kitchen and toilet	1 per meal served	0.2
Dining hall—kitchen and toilet waste with dishwasher or food waste grinder or both	1 per meal served	0.25
Dining hall—kitchen only	1 per meal served	0.06
Drive-in restaurant, inside seating	1 per seat	0.3
Drive-in restaurant, without inside seating	1 per car space	0.3
Drive-in theater	1 per car space	0.1
Employees—in all buildings	1 per person	0.4
Floor drain	1 per drain	1.0
Hospital	1 per bed space	2.0
Hotel or motel and tourist rooming house	1 per room	0.9
Labor camp—central bathhouse	1 per employee	0.25
Medical office buildings, clinics and dental offices		
Doctors, nurses and medical staff	1 per person	0.8
Office personnel	1 per person	0.25
Patients	1 per person	0.15
Mobile home park	1 per mobile home site	3.0
Motor-fuel-dispensing facility	1 per car served	0.15

TABLE 604.1(2)—CONVERSION FACTOR—continued		
BUILDING CLASSIFICATION	**UNITS**	**FACTOR**
Nursing or group homes	1 per bed space	1.0
Outdoor sports facility—toilet waste only	1 per person	0.35
Park—showers and toilets	1 per acre	8.0
Park—toilet waste only	1 per acre	4.0
Restaurant—dishwasher or food waste grinder or both	1 per seating space	0.15
Restaurant—kitchen and toilet	1 per seating space	0.6
Restaurant—kitchen waste only	1 per seating space	0.18
Restaurant—toilet waste only	1 per seating space	0.42
Restaurant—(24-hour) kitchen and toilet	1 per seating space	1.2
Restaurant—(24-hour) with dishwasher or food waste grinder or both	1 per seating space	1.5
Retail store	1 per customer	0.03
School—meals and showers	1 per classroom	8.0
School—meals served or showers	1 per classroom	6.7
School—no meals, no showers	1 per classroom	5.0
Self-service laundry—toilet waste only	1 per machine	1.0
Showers—public	1 per shower	0.3
Swimming pool bathhouse	1 per person	0.2

SECTION 605—INSTALLATION OF CONVENTIONAL SOIL ABSORPTION SYSTEMS

605.1 Seepage trench excavations. *Seepage trench* excavations shall be 1 foot to 5 feet (305 mm to 1524 mm) wide. Trench excavations shall be spaced not less than 6 feet (1829 mm) apart. The absorption area of a *seepage trench* shall be computed by using only the bottom of the trench area. The bottom excavation area of the distribution header shall not be computed as absorption area. Individual *seepage trenches* shall be not greater than 100 feet (30 480 mm) long, except as otherwise approved.

605.2 Seepage bed excavations. *Seepage bed* excavations shall be not less than 5 feet (1524 mm) wide and have more than one distribution pipe. The absorption area of a *seepage bed* shall be computed by using the bottom of the trench area. Distribution piping in a *seepage bed* shall be uniformly spaced not greater than 5 feet (1524 mm) and not less than 3 feet (914 mm) apart, and not greater than 3 feet (914 mm) and not less than 1 foot (305 mm) from the sidewall or headwall.

605.3 Seepage pits. A *seepage pit* shall have not less than an inside diameter of 5 feet (1524 mm) and shall consist of a chamber walled-up with material, such as perforated precast concrete ring, concrete block, brick or other approved material allowing effluent to percolate into the surrounding soil. The pit bottom shall be left open to the soil. Aggregate of $^1/_2$ inch to $2^1/_2$ inches (12.7 mm to 64 mm) in size shall be placed into a 6-inch minimum (152 mm) annular space separating the outside wall of the chamber and sidewall excavation. The depth of the annular space shall be measured from the inlet pipe to the bottom of the chamber. Each *seepage pit* shall be provided with a 24-inch (610 mm) manhole extending to within 56 inches (1422 mm) of the ground surface and a 4-inch-diameter (102 mm) fresh air inlet. *Seepage pits* shall be located not less than 5 feet (1524 mm) apart. Excavation and scarifying shall be in accordance with Section 605.4. The effective area of a *seepage pit* shall be the vertical wall area of the walled-up chamber for the depth below the inlet for all strata in which the percolation rates are less than 30 minutes per inch (70 s/mm). The 6-inch (152 mm) annular opening outside the vertical wall area is permitted to be included for determining the effective area. Table 605.3 or an approved method shall be used for determining the effective sidewall area of circular *seepage pits*.

TABLE 605.3—EFFECTIVE SQUARE-FOOT ABSORPTION AREA FOR SEEPAGE PITS						
INSIDE DIAMETER OF CHAMBER IN FEET PLUS 1 FOOT FOR WALL THICKNESS PLUS 1 FOOT FOR ANNULAR SPACE	**DEPTH IN FEET OF PERMEABLE STRATA BELOW INLET**					
	3	**4**	**5**	**6**	**7**	**8**
7	47	88	110	132	154	176
8	75	101	126	151	176	201
9	85	113	142	170	198	226
10	94	126	157	188	220	251
11	104	138	173	208	242	277
13	123	163	204	245	286	327
For SI: 1 foot = 304.8 mm.						

605.4 Excavation and construction. The bottom of a trench or bed excavation shall be level. *Seepage trenches* or beds shall not be excavated where the soil is so wet that such material rolled between the hands forms a soil wire. Smeared or compacted soil surfaces in the sidewalls or bottom of *seepage trench* or bed excavations shall be scarified to the depth of smearing or compaction and the loose material removed. Where rain falls on an open excavation, the soil shall be left until sufficiently dry so a soil wire will not form when soil from the excavation bottom is rolled between the hands. The bottom area shall then be scarified and loose material removed.

605.5 Aggregate and backfill. Not less than 6 inches (152 mm) of aggregate ranging in size from $^1/_2$ inch to $2^1/_2$ inches (12.7 mm to 64 mm) shall be laid into the trench or bed below the distribution pipe elevation. The aggregate shall be evenly distributed not less than 2 inches (51 mm) over the top of the distribution pipe. The aggregate shall be covered with approved synthetic materials or 9 inches (229 mm) of uncompacted marsh hay or straw. Building paper shall not be used to cover the aggregate. Not less than 18 inches (457 mm) of soil backfill shall be provided above the covering.

605.6 Distribution piping. Distribution piping for gravity systems shall be not less than 4 inches (102 mm) in diameter. The distribution header (PVC) shall be solid-wall pipe. The top of the distribution pipe shall be not less than 8 inches (203 mm) below the original surface in continuous straight or curved lines. The slope of the distribution pipes shall be 2 inches to 4 inches (51 mm to 102 mm) per 100 feet (30 480 mm). Effluent shall be distributed to all distribution pipes. Distribution of effluent to *seepage trenches* on sloping sites shall be accomplished by using a drop box design or other approved methods. Where dosing is required, the siphon or pump shall discharge a dose of minimum capacity equal to 75 percent of the combined volume of the distribution piping in the absorption system.

605.7 Observation pipes. Observation pipes shall be provided. Such pipes shall be not less than 4 inches (102 mm) in diameter, not less than 12 inches (305 mm) above final grade and shall terminate with an approved vent cap.

The bottom 12 inches (305 mm) of the observation pipe shall be perforated and extend to the bottom of the aggregate. Observation pipes shall be located not less than 25 feet (7620 mm) from any window, door or air intake of any building used for human occupancy. Not more than four distribution pipelines shall be served by one common 4-inch (102 mm) observation pipe where interconnected by a common header pipe (see Appendix-A, Figure A101.1(4)).

Exception: Where approved and where the location of the observation pipe is permanently recorded, the observation pipe shall be not more than 2 inches (51 mm) below the finished grade.

605.8 Winter installation. Soil absorption systems shall not be installed during periods of adverse weather conditions unless the installation is approved. A soil absorption system shall not be installed where the soil at the system elevation is frozen. Snow cover shall be removed from the soil absorption area before excavation begins. Snow shall not be placed in a manner that will cause water to pond on the soil absorption system area during snow melt. Excavated soil to be used as backfill shall be protected from freezing. Excavated soil that freezes solid shall not be used as backfill. The first 12 inches (305 mm) of backfill shall be loose, unfrozen soil. Inspection of systems installed during winter conditions shall include inspection of the trench or bed excavation prior to the placement of gravel and inspection of backfill material at the time of placement.

605.9 Evaporation. Soil absorption systems shall not be covered or paved over by material that inhibits the evaporation of the effluent.

CHAPTER 7 PRESSURE DISTRIBUTION SYSTEMS

User notes:

About this chapter: *Chapter 7 regulates the pressure distribution method of soil absorption systems. This type of system is necessary where the water from the outlet of a septic tank cannot flow by gravity because of site constraints. The water is pumped from a collection tank to the absorption area at regular intervals.*

SECTION 701—GENERAL

701.1 Scope. The provisions of this chapter shall govern the design and installation of *pressure distribution systems.*

SECTION 702—DESIGN LOADING RATE

702.1 General. A *pressure distribution system* shall be permitted for use on any site meeting the conventional *private sewage disposal system* criteria. There shall be not less than 6 inches (152 mm) to the top of the distribution piping from original grade for any *pressure distribution system.* The minimum required suitable soil depths from original grade for *pressure distribution systems* shall be in accordance with Table 702.1.

TABLE 702.1—SOIL REQUIRED

DISTRIBUTION PIPE (inches)	SUITABLE SOIL (inches)
1	49
2	50
3	52
4	53

For SI: 1 inch = 25.4 mm.

702.2 Absorption area. The total absorption area required shall be computed from the estimated daily wastewater flow and the design loading rate based on the percolation rate for the site. The required absorption area equals wastewater flow divided by the design loading rate from Table 702.2. Two systems of equal size shall be required for systems receiving effluents exceeding 5,000 gallons (18 925 L). Each system shall have a minimum capacity of 75 percent of the area required for a single system and shall be provided with a suitable means of alternating waste applications. A dual system shall be considered as one system.

TABLE 702.2—DESIGN LOADING RATE

PERCOLATION RATE (minutes per inch)	DESIGN LOADING FACTOR (gallons per square foot per day)
0 to less than 10	1.2
10 to less than 30	0.8
30 to less than 45	0.72
45 to 60	0.4

For SI: 1 minute per inch = 2.4 s/mm, 1 gallon per square foot = 0.025 L/m^2.

702.3 Estimated wastewater flow. The estimated wastewater flow from a residence shall be 150 gallons (568 L) per bedroom per day. Wastewater flow rates for other occupancies in a 24-hour period shall be based on the values in Table 802.7.2.

SECTION 703—SYSTEM DESIGN

703.1 General. *Pressure distribution systems* shall discharge effluent into trenches or beds. Each pipe connected to an outlet of a manifold shall be counted as a separate distribution pipe. The horizontal spacing of distribution pipes shall be 30 inches to 72 inches (762 mm to 1829 mm). The system shall be sized in accordance with the formulas listed in this section. Systems using Schedule 40 plastic pipe shall be sized in accordance with the formulas listed in this section or in accordance with the tables listed in Appendix B. Distribution piping shall be installed at the same elevation, unless an approved system provides for a design ensuring equal flow through each of the perforations and the effluent is uniformly applied to the soil infiltrative surface [see Appendix A, Figure A101.1(5)].

703.2 Symbols. The following symbols and notations shall apply to the provisions of this chapter:

C_h = Hazen-Williams friction factor.

D = Distribution pipe diameter, inches (mm).

d = Perforation diameter, inches (mm).

D_d = Delivery pipe diameter, inches (mm).

D_m = Manifold pipe diameter, inches (mm).

f = Fraction of total head loss in the manifold segment.

F_D = Friction loss in the delivery pipe, feet of head (mm of head).

F_i = Friction factor for i^{th} manifold segment.

F_N = Friction loss in the network pipe, feet of head (mm of head).

h = Pressure in distribution pipe, feet of head (mm of head).

h_d = In-line pressure at distal end of lateral, feet of head (mm of head).

L_D = Length of delivery pipe, feet (mm).

L_i = Length of ith manifold segment, feet (mm).

N = Number of perforations in the lateral.

q = Perforation discharge rate, gpm (L/min).

Q_i = Flow rate i^{th} manifold segment, gpm (L/min).

Q_m = Flow rate at manifold inlet, gpm (L/min).

703.3 Distribution pipe. Distribution pipe size, hole diameter and hole spacing shall be selected. The hole diameter and spacing shall be equal for each manifold segment. Distribution pipe size shall not be required to be the same for each segment. Changes in pressure in the distribution pipe shall be less than or equal to 10 percent by conforming to the following formula:

Formula 7-1 $\sum \Delta h 0.2 h_d$

For SI: 1 foot = 304.8 mm.

where:

$$\Delta h = 4.71L\left(\frac{q}{C_h D^{2.65}}\right)^{1.85}$$

$$q = 11.79d^2\sqrt{h_d}$$

The Hazen-Williams friction factor, C_h, for each pipe material shall be determined in accordance with Table 703.3.

TABLE 703.3—HAZEN-WILLIAMS FRICTION FACTOR

MATERIAL	FRICTION FACTOR, C_h
ABS plastic pipe	150
Asbestos-cement pipe	140
Bituminized fiber pipe	120
Cast-iron pipe	100
Concrete pipe	110
Copper or copper-alloy tubing	150
PVC plastic pipe	150
Vitrified clay pipe	100

703.4 Manifolds. The diameter of the manifold pipe shall be determined by the following equation:

Equation 7-1 $$D_m = \left(\frac{\sum L_i F_i}{f h_d}\right)^{0.21}$$

For SI: 1 inch = 25.4 mm.

where:

$F_i = 9.8 \times 10^{-4} Q_i$

$q = 11.79 d^2 \sqrt{h_d}$

$Q = Nq$

The fraction of the total head loss at the manifold segment, *f*, shall be less than or equal to 0.1. The in-line pressure at the distal end of the lateral, h_d, shall be not less than 2.5 feet (762 mm) of head. Distribution pipes shall be connected to the manifold with tees or 90-degree (1.57 rad) ells. Distribution pipes shall have the ends capped.

703.5 Friction loss. The delivery pipe shall include all pipe between the pump and the supply end of the distribution pipe. The friction loss in the delivery pipe, F_D, shall be determined by the following equation:

Equation 7-2 $$F_D = L_D\left(\frac{3.55 Q_m}{C_h D_d^{2.63}}\right)^{1.85}$$

For SI: 1 inch of head = 25.4 mm of head.

The Hazen-Williams friction factor, C_h, for each pipe material shall be determined in accordance with Table 703.3.

The friction loss in the network pipe shall be determined by the following equation:

Equation 7-3 $F_N = 1.31\, h_d$

For SI: 1 inch of head = 25.4 mm of head.

Pipe in the system shall be increased in size if the friction loss is excessive.

703.6 Force main. Size of the force main between the pump and manifold shall be based on the friction loss and velocity of effluent through the pipe. The velocity of effluent in a force main shall be not more than 5 feet per second (1524 mm/sec).

SECTION 704—BED AND TRENCH CONSTRUCTION

704.1 General. The excavation and construction for *pressure distribution system* trenches and beds shall be in accordance with Chapter 6. Aggregate shall be not less than 6 inches (152 mm) beneath the distribution pipe with 2 inches (51 mm) spread evenly above the pipe. The aggregate shall be clean, nondeteriorating 0.5-inch to 2.5-inch (12.7 mm to 64 mm) stone.

SECTION 705—PUMPS

705.1 General. Pump selection shall be based on the discharge rate and total dynamic head of the pump performance curve. The total dynamic head shall be equal to the difference in feet (mm) of elevation between the pump and distribution pipe invert plus the friction loss and not less than 2.5 feet (762 mm) where using low pressure distribution in the delivery pipe and network pipe.

705.2 Pump and alarm controls. The control system for the pumping chamber shall consist of a control for operating the pump and an alarm system to detect a pump. Pump start and stop depth controls shall be adjustable. Pump and alarm controls shall be of an approved type. Switches shall be resistant to sewage corrosion.

705.3 Alarm system. Alarm systems shall consist of a bell or light, mounted in the structure, and shall be located to be easily seen or heard. The high-water sensing device shall be installed approximately 2 inches (51 mm) above the depth set for the "on" pump control but below the bottom of the inlet to the pumping chamber. Alarm systems shall be installed on a separate circuit from the electrical service.

705.4 Electrical connections. Electrical connections shall be located outside the pumping chamber.

SECTION 706—DOSING

706.1 General. The dosing frequency shall be not greater than four times daily. A volume per dose shall be established by dividing the daily wastewater flow by the dosing frequency. The dosing volume shall be not less than 10 times the capacity of the distribution pipe volume. Table 706.1 provides the estimated volume for various pipe diameters.

TABLE 706.1—ESTIMATED VOLUME FOR VARIOUS DIAMETER PIPES

DIAMETER (inches)	VOLUME (gallons per foot length)
1	0.041
$1^1/_4$	0.064
$1^1/_2$	0.092
2	0.164
3	0.368
4	0.655
5	1.47

For SI: 1 inch = 25.4 mm, 1 gallon per foot = 0.012 L/mm.

CHAPTER

8 TANKS

User notes:

About this chapter: *Septic tanks and other treatment tanks are key components of private sewage disposal systems as they must be properly sized to achieve the desired reduction of sewage to its basic components of sludge and nonpotable water. Chapter 8 covers sizing, capacity and installation of these tanks and water holding tanks.*

SECTION 801—GENERAL

801.1 Scope. The provisions of this chapter shall govern the design, installation, repair and maintenance of septic tanks, treatment tanks and holding tanks.

SECTION 802—SEPTIC TANKS AND OTHER TREATMENT TANKS

802.1 General. Septic tanks shall be fabricated or constructed of welded steel, monolithic concrete, fiberglass or an approved material. Tanks shall be watertight and fabricated to constitute an individual structure, and shall be designed and constructed to withstand anticipated loads. The design of prefabricated septic tanks shall be approved. Plans for site-constructed concrete tanks shall be approved prior to construction.

802.2 Design of septic tanks. Septic tanks shall have not less than two compartments. The inlet compartment shall be not less than two-thirds of the total capacity of the tank, not less than a 500-gallon (1893 L) liquid capacity and not less than 3 feet (914 mm) wide and 5 feet (1524 mm) long. The secondary compartment of a septic tank shall have not less than a capacity of 250 gallons (946 L) and not more than one-third of the total capacity. The secondary compartment of septic tanks having a capacity more than 1,500 gallons (5678 L) shall be not less than 5 feet (1524 mm) long.

The liquid depth shall be not less than 30 inches (762 mm) and a maximum average of 6 feet (1829 mm). The total depth shall be not less than 8 inches (203 mm) greater than the liquid depth.

Rectangular tanks shall be constructed with the longest dimensions parallel to the direction of the flow.

Cylindrical tanks shall be not less than 48 inches (1219 mm) in diameter.

802.3 Inlets and outlets. The inlet and outlet on all tanks or tank compartments shall be provided with open-end coated sanitary tees or baffles made of approved materials constructed to distribute flow and retain scum in the tank or compartments. The inlet and outlet openings on all tanks shall contain a stop or other provision that will prevent the insertion of the sewer piping beyond the inside wall of the tank. The tees or baffles shall extend not less than 6 inches (152 mm) above the liquid level, not less than 9 inches (229 mm) below the liquid level, and not greater than one-third the liquid depth below the liquid level. Not less than 2 inches (51 mm) of clear space shall be provided above the top of the baffles or tees. The bottom of the outlet opening shall be not less than 2 inches (51 mm) below the bottom of the inlet.

802.4 Manholes. Each compartment of a tank shall be provided with not fewer than one manhole opening located over the inlet or outlet opening, and such opening shall be not less than 24 inches (610 mm) square or 24 inches (610 mm) in diameter. Where the inlet compartment of a septic tank exceeds 12 feet (3658 mm) in length, an additional manhole shall be provided over the baffle wall. Manholes shall terminate not greater than 6 inches (152 mm) below the ground surface. Steel tanks shall have not less than a 2-inch (51 mm) collar for the manhole extensions permanently welded to the tank. The manhole extension on fiberglass tanks shall be of the same material as the tank and an integral part of the tank. The collar shall be not less than 2 inches (51 mm) high.

802.5 Manhole covers. Manhole risers shall be provided with a fitted, water-tight cover of concrete, steel, cast iron or other approved material capable of withstanding all anticipated loads. Manhole covers terminating above grade shall have an approved locking device.

802.6 Inspection opening. An inspection opening shall be provided over either the inlet or outlet baffle of every treatment tank. The opening shall be not less than 4 inches (102 mm) in diameter with a tight-fitting cover. Inspection pipes terminating above ground shall be not less than 6 inches (152 mm) above finished grade. Inspection pipes approved for terminating below grade shall be not more than 2 inches (51 mm) below finished grade, and the location shall be permanently recorded.

802.7 Capacity and sizing. The capacity of a septic tank or other treatment tank shall be based on the number of persons using the building to be served or on the volume and type of waste, whichever is greater. The minimum liquid capacity shall be 750 gallons (2839 L). Where the required capacity is to be provided by more than one tank, the minimum capacity of any tank shall be 750 gallons (2839 L). The installation of more than four tanks in series is prohibited.

802.7.1 Sizing of tank. The minimum liquid capacity for one- and two-family dwellings shall be in accordance with Table 802.7.1.

TABLE 802.7.1—SEPTIC TANK CAPACITY FOR ONE- AND TWO-FAMILY DWELLINGS

NUMBER OF BEDROOMS	SEPTIC TANK (gallons)
1	750
2	750
3	1,000
4	1,200
5	1,425
6	1,650
7	1,875
8	2,100

For SI: 1 gallon = 3.785 L.

802.7.2 Other buildings. For buildings, the liquid capacity shall be increased above the 750-gallon (2839 L) minimum as established in Table 802.7.1. In buildings with kitchen or laundry waste, the tank capacity shall be increased to receive the anticipated volume for a 24-hour period from the kitchen or laundry or both. The liquid capacities established in Table 802.7.2 do not include employees.

Exception: One- or two-family dwellings.

TABLE 802.7.2—ADDITIONAL CAPACITY FOR OTHER BUILDINGS

BUILDING CLASSIFICATION	CAPACITY (gallons)
Apartment buildings (per bedroom—includes automatic clothes washer)	150
Assembly halls (per person—no kitchen)	2
Bars and cocktail lounges (per patron space)	9
Beauty salons (per station—includes customers)	140
Bowling centers (per lane)	125
Bowling centers with bar (per lane)	225
Camp, day use only—no meals served (per person)	15
Campgrounds and camping resorts (per camp space)	100
Campground sanitary dump stations (per camp space) (omit camp spaces with sewer connection)	5
Camps, day and night (per person)	40
Car washes (per car handwash)	50
Catch basins—such as for garages and motor-fuel-dispensing facilities (per basin)	100
Catch basins—truck washing (per truck)	100
Places of religious worship—no kitchen (per person)	3
Places of religious worship—with kitchen (per person)	7.5
Condominiums (per bedroom—includes automatic clothes washer)	150
Dance halls (per person)	3
Dining halls—kitchen and toilet waste—with dishwasher, food waste grinder or both (per meal served)	11
Dining halls—kitchen waste only (per meal served)	3
Drive-in restaurants—all paper service (per car space)	15
Drive-in restaurants—all paper service, inside seating (per seat)	15
Drive-in theaters (per car space)	5
Employees—in all buildings, per employee—total all shifts	20
Floor drains (per drain)	50
Hospitals (per bed space)	200
Hotels or motels and tourist rooming houses	100
Labor camps, central bathhouses (per employee)	30

TABLE 802.7.2—ADDITIONAL CAPACITY FOR OTHER BUILDINGS—continued	
BUILDING CLASSIFICATION	**CAPACITY (gallons)**
Medical office buildings, clinics and dental offices Doctors, nurses, medical staff (per person) Office personnel (per person) Patients (per person)	 75 20 10
Mobile home parks, homes with bathroom groups (per site)	300
Motor-fuel-dispensing facilities	10
Nursing and rest homes—without laundry (per bed space)	100
Outdoor sports facilities (toilet waste only—per person)	5
Parks, toilet waste (per person—75 persons per acre)	5
Parks, with showers and toilet waste (per person—75 persons per acre)	10
Restaurants—dishwasher or food waste grinder or both (per seat)	3
Restaurants—kitchen and toilet wastes (per seating space)	30
Restaurants—kitchen waste only—without dishwasher and food waste grinder (per seat)	9
Restaurants—toilet waste only (per seat)	21
Restaurants (24-hour)—dishwasher or food waste grinder (per seat)	6
Restaurants (24-hour)—kitchen and toilet waste (per seating space)	60
Retail stores—customers	1.5
Schools (per classroom—25 pupils per classroom)	450
Schools with meals served (per classroom—25 pupils per classroom)	600
Schools with meals served and showers provided (per classroom)	750
Self-service laundries (toilet waste only, per machine) Automatic clothes washers—such as for apartments and service buildings, (per machine)	50 300
Showers—public (per shower taken)	15
Swimming pool bathhouses (per person)	10
For SI: 1 gallon = 3.785 L.	

802.8 Installation. Septic and other treatment tanks shall be located with a horizontal distance not less than specified in Table 802.8 between various elements. Tanks installed in ground water shall be securely anchored. A 3-inch-thick (76 mm) compacted bedding shall be provided for all septic and other treatment tank installations. The bedding material shall be sand, gravel, granite, limerock or other noncorrosive materials of such size that the material passes through a 0.5-inch (12.7 mm) screen.

TABLE 802.8—MINIMUM HORIZONTAL SEPARATION DISTANCES FOR TREATMENT TANKS	
ELEMENT	**DISTANCE (feet)**
Building	5
Cistern	25
Foundation wall	5
Lake, high water mark	25
Lot line	2
Pond	25
Reservoir	25
Spring	50
Stream or watercourse	25
Swimming pool	15
Water service	5
Well	25
For SI: 1 foot = 304.8 mm.	

802.9 Backfill. The backfill material for steel and fiberglass tanks shall be specified for bedding and shall be tamped into place without causing damage to the coating. The backfill for concrete tanks shall be soil material, which shall pass a 4-inch (102 mm) screen and be tamped into place.

802.10 Manhole riser joints. Joints on concrete risers and manhole covers shall be tongue-and-groove or shiplap type and sealed water tight using neat cement, mortar or bituminous compound. Joints on steel risers shall be welded or flanged and bolted and water tight. Steel manhole extensions shall be bituminous coated both inside and outside. Methods of attaching fiberglass risers shall be water tight and approved.

802.11 Dosing or pumping chambers. Dosing or pumping chambers shall be fabricated or constructed of welded steel, monolithic concrete, glass fiber-reinforced polyester or other approved materials. Manholes for dosing or pumping chambers shall terminate not less than 4 inches (102 mm) above the ground surface. Dosing or pumping chambers shall be water tight, and materials and construction specifications shall meet the same criteria specified for septic tanks in this chapter.

802.11.1 Capacity sizing. The working capacity of the dosing or pumping chamber shall be sized to permit automatic discharge of the total daily sewage flow with discharge occurring not more than four times per 24 hours. Minimum capacity of a dosing chamber shall be 500 gallons (1893 L) and a space shall be provided between the bottom of the pump and floor of the dosing or pumping chamber. A dosing chamber shall have a 1-day holding capacity located above the high-water alarm for one- and two-family dwellings based on 100 gallons (379 L) per day per bedroom, or in the case of other buildings, in accordance with Section 802.7. Minimum pump chamber sizes are indicated for one- and two-family dwellings in Table 802.11.1. Where the total developed length of distribution piping exceeds 1,000 feet (305 m), the dosing or pumping chamber shall have two siphons or pumps dosing alternately and serving one-half of the soil absorption system.

TABLE 802.11.1—PUMP CHAMBER SIZES

NUMBER OF BEDROOMS	MINIMUM PUMPING CHAMBER SIZE (gallons)
1	500
2	500
3	750
4	750
5	1,000

For SI: 1 gallon = 3.785 L.

802.12 Design of other treatment tanks. The design of other treatment tanks shall be approved on an individual basis. The capacity, sizing and installation of the tank shall be in accordance with this section except as otherwise approved. Where a treatment tank is preceded by a conventional septic tank, credit shall be given for the capacity of the septic tank.

SECTION 803—MAINTENANCE AND SLUDGE DISPOSAL

803.1 Maintenance. Septic tanks and other treatment tanks shall be cleaned whenever the sludge and scum occupy one-third of the tank's liquid capacity.

803.2 Septage. Septage shall be disposed of at an approved location.

SECTION 804—CHEMICAL RESTORATION

804.1 General. Products for chemical restoration or chemical restoration procedures for *private sewage disposal systems* shall not be used unless approved.

SECTION 805—HOLDING TANKS

805.1 Approval. The installation of a holding tank shall not be approved where the site can accommodate the installation of any other *private sewage disposal system* specified in this code. A pumping and maintenance schedule for each holding tank installation shall be submitted to the *code official.*

805.2 Sizing. The minimum liquid capacity of a holding tank for one- and two-family dwellings shall be in accordance with Table 805.2. Other buildings shall have a minimum 5-day holding capacity, but not less than 2,000 gallons (7570 L). Sizing shall be in accordance with Table 802.7.2. Not more than four holding tanks shall be installed in series.

TABLE 805.2—MINIMUM LIQUID CAPACITY OF HOLDING TANKS

NUMBER OF BEDROOMS	TANK CAPACITY (gallons)
1	2,000
2	2,000

TABLE 805.2—MINIMUM LIQUID CAPACITY OF HOLDING TANKS—continued	
NUMBER OF BEDROOMS	TANK CAPACITY (gallons)
3	2,000
4	2,500
5	3,000
6	3,500
7	4,000
8	4,500
For SI: 1 gallon = 3.785 L.	

805.3 Construction. Holding tanks shall be constructed of welded steel, monolithic concrete, glass-fiber-reinforced polyester or other approved materials.

805.4 Installation. Tanks shall be located in accordance with Section 802.8, except the tanks shall be not less than 20 feet (6096 mm) from any part of a building. Holding tanks shall be located so the servicing manhole is located not less than 10 feet (3048 mm) from an all-weather access road or drive.

805.5 Warning device. A high-water warning device shall be installed to activate 1 foot (305 mm) below the inlet pipe. This device shall be either an audible or an approved illuminated alarm. The electrical junction box, including warning equipment junctions, shall be located outside the holding tank or housed in waterproof, explosionproof enclosures. Electrical relays or controls shall be located outside the holding tank.

805.6 Manholes. Each tank shall be provided with either a manhole not less than 24 inches (610 mm) square or with a manhole having a 24-inch (610 mm) inside diameter extending not less than 4 inches (102 mm) above ground. Finished grade shall be sloped away from the manhole to divert surface water from the manhole. Each manhole cover shall have an effective locking device. Service ports in manhole covers shall be not less than 8 inches (203 mm) in diameter and shall be 4 inches (102 mm) above finished grade level. The service port shall have an effective locking cover or a brass cleanout plug.

805.7 Septic tank. The outlet shall be sealed where an approved septic tank is installed to serve as a holding tank. Removal of the inlet and outlet baffle shall not be prohibited.

805.8 Vent. Each tank shall be provided with a vent not less than 2 inches (51 mm) in diameter and shall extend not less than 12 inches (305 mm) above finished grade, terminating with a return bend fitting or approved vent cap.

CHAPTER 9 MOUND SYSTEMS

User notes:

About this chapter: *Mound systems are another method for disposal of the nonpotable water from a septic tank. The existing soil at the site may not have the capacity to absorb the water because of seasonal ground water conditions or presence of bedrock just below the soil surface. Chapter 9 provides extensive design information and installation methods to build mound systems for absorbing the nonpotable water.*

SECTION 901—GENERAL

901.1 Scope. The provisions of this chapter shall govern the design and installation of mound systems.

SECTION 902 —SOIL AND SITE REQUIREMENTS

902.1 Soil borings. Not less than three *soil borings* per site shall be conducted in accordance with Chapter 4 to determine the depth to seasonal or permanent soil saturation or bedrock. Identification of a replacement system area is not required.

902.2 Prohibited locations. A mound system shall be prohibited on sites not having the minimum depths of soil specified in Table 902.2. The installation of a mound in a filled area shall be prohibited. A mound shall not be installed in a compacted area or over a failing conventional system.

TABLE 902.2—MINIMUM SOIL DEPTHS FOR MOUND SYSTEM INSTALLATION

RESTRICTING FACTOR	MINIMUM SOIL DEPTH TO RESTRICTION (inches)
High ground water	24
Impermeable rock strata	60
Pervious rock	24
Rock fragments (50-percent volume)	24

For SI: 1 inch = 25.4 mm.

902.3 Slowly permeable soils with or without high ground water. Percolation tests shall be conducted at a depth of 20 inches to 24 inches (508 mm to 610 mm) from existing grade. Where a more slowly permeable horizon exists at less than 20 inches to 24 inches (508 mm to 610 mm), percolation tests shall be conducted within that horizon. A mound system shall be suitable for such site condition where the percolation rate is greater than 60 minutes per inch and less than or equal to 120 minutes per inch (2.4 min/mm to 4.7 min/mm).

902.4 Shallow permeable soils over creviced bedrock. Percolation tests shall be conducted at a depth of 12 inches to 18 inches (305 mm to 457 mm) from existing grade. Where a more slowly permeable horizon exists within 12 inches to 18 inches (305 mm to 457 mm), percolation tests shall be conducted within that horizon. A mound system shall be suitable for such site condition where the percolation rate is between 3 minutes per inch and 60 minutes per inch (0.12 min/mm and 2.4 min/mm).

902.5 Permeable soils with high ground water. Percolation tests shall be conducted at a depth of 20 inches to 24 inches (508 mm to 610 mm) from existing grade. Where a more slowly permeable horizon exists at less than 20 inches to 24 inches (508 mm to 610 mm), percolation tests shall be conducted within that horizon. A mound system shall be suitable for such site condition where the percolation rate is between 0 minutes per inch and 60 minutes per inch (0 min/mm and 2.4 min/mm).

902.6 Depth to pervious rock. Not less than 24 inches (610 mm) of unsaturated natural soil shall be over creviced or porous bedrock.

902.7 Depth to high ground water. Not less than 24 inches (610 mm) of unsaturated natural soil shall be present over high ground water as indicated by soil mottling or direct observation of water in accordance with Chapter 4.

902.8 Slopes. A mound shall not be installed on a slope greater than 6 percent where the percolation rate is between 30 and 120 minutes per inch (1.2 and 4.7 min/mm). The maximum allowable slope shall be 12 percent where there is a complex slope (two directions).

902.9 Location of mound on sloping sites. The mound shall be located so the longest dimension of the mound and the distribution lines are perpendicular to the slope. The mound shall be placed upslope and not at the base of a slope. The mound shall be situated so the effluent is not concentrated in one direction where there is a complex slope (two directions). Surface water runoff shall be diverted around the mound.

902.10 Depth to rock strata or 50 percent by volume rock fragments. Not less than 60 inches (1524 mm) of soil shall be present over uncreviced, impermeable bedrock. Where the soil contains 50-percent coarse fragments by volume in the upper 24 inches (610 mm), a mound shall not be installed except where there is not less than 24 inches (610 mm) of permeable, unsaturated soil with less than 50-percent coarse fragments located beneath this layer.

SECTION 903—SYSTEM DESIGN

903.1 Mound dimensions and design. For one- and two-family dwellings and other buildings with estimated wastewater flows less than 600 gallons (2271 L) per day, the mound dimensions shall be determined in accordance with this section or Tables 903.1(1) through 903.1(12). Dimensions and corresponding letter designations listed in the tables and referenced in this section are shown in Appendix A, Figures A101.1(6) through A101.1(10). For buildings with estimated waste-water flows exceeding 600 gallons (2271 L) per day, the mound shall be designed in accordance with this section. Daily wastewater flow shall be estimated as 150 gallons (568 L) per day per bedroom for one- and two-family dwellings. For other buildings the total daily wastewater flow shall be determined in accordance with Table 802.7.2.

TABLE 903.1(1)—DESIGN CRITERIA FOR A MOUND FOR A ONE-BEDROOM HOME ON A 0- TO 6-PERCENT SLOPE WITH LOADING RATES OF 150 GALLONS PER DAY FOR SLOWLY PERMEABLE SOIL

DESIGN PARAMETER		SLOPE (percent)			
		0	2	4	6
A	Trench width, feet	3	3	3	3
B	Trench length, feet Number of trenches	42 1	42 1	42 1	42 1
D	Mound height, inches	12	12	12	12
F	Mound height, inches	9	9	9	9
G	Mound height, inches	12	12	12	12
H	Mound height, inches	18	18	18	18
I	Mound width, feet[a]	15	15	15	15
J	Mound width, feet[a]	11	8	8	8
K	Mound length, feet	10	10	10	10
L	Mound length, feet	62	62	62	62
P	Distribution pipe length, feet Distribution pipe diameter, inches Number of holes per distribution pipe[b] Hole spacing, inches[b] Hole diameter, inches[b]	20 1 9 30 0.25	20 1 9 30 0.25	20 1 9 30 0.25	20 1 9 30 0.25
W	Mound width, feet	25	26	26	26

For SI: 1 inch = 25.4 mm, 1 foot = 304.8 mm, 1 gallon = 3.785 L.
a. Additional width to obtain required basal area.
b. Last hole is located at the end of the distribution pipe, which is 15 inches from the other hole.

TABLE 903.1(2)—DESIGN CRITERIA FOR A TWO-BEDROOM HOME FOR A MOUND ON A 0- TO 6-PERCENT SLOPE WITH LOADING RATES OF 300 GALLONS PER DAY FOR SLOWLY PERMEABLE SOIL

DESIGN PARAMETER		SLOPE (percent)			
		0	2	4	6
A	Trench width, feet	3	3	3	3
B	Trench length, feet Number of trenches	42 2	42 2	42 2	42 2
C	Trench spacing, feet	15	15	15	15
D	Mound height, inches	12	12	12	12
E	Mound height, inches	12	17	25	25
F	Mound height, inches	9	9	9	9
G	Mound height, inches	12	12	12	12
H	Mound height, inches	18	18	18	18
I	Mound width, feet[a]	12	20	20	20
J	Mound width, feet	12	8	8	8

TABLE 903.1(2)—DESIGN CRITERIA FOR A TWO-BEDROOM HOME FOR A MOUND ON A 0- TO 6-PERCENT SLOPE WITH LOADING RATES OF 300 GALLONS PER DAY FOR SLOWLY PERMEABLE SOIL—continued

DESIGN PARAMETER		SLOPE (percent)			
		0	2	4	6
K	Mound length, feet	10	10	10	10
L	Mound length, feet	62	62	62	62
P	Distribution pipe length, feet Distribution pipe diameter, inches Number of holes per distribution pipe[b] Hole spacing, inches[b] Hole diameter, inches	20 1 9 30 0.25	20 1 9 30 0.25	20 1 9 30 0.25	20 1 9 30 0.25
R	Manifold length, feet Manifold diameter, inches[c]	1 52	1 52	1 52	1 52
W	Mound width, feet	42	46	46	46

For SI: 1 inch = 25.4 mm, 1 foot = 304.8 mm, 1 gallon = 3.785 L.

a. Additional width to obtain required basal area.

b. Last hole is located at the end of the distribution pipe, which is 15 inches from the other hole.

c. Diameter dependent on the size of pipe from pump and inlet position.

TABLE 903.1(3)—DESIGN CRITERIA FOR A THREE-BEDROOM HOME FOR A MOUND ON A 0- TO 6-PERCENT SLOPE WITH LOADING RATES OF 450 GALLONS PER DAY FOR SLOWLY PERMEABLE SOIL

DESIGN PARAMETER		SLOPE (percent)			
		0	2	4	6
A	Trench width, feet	3	3	3	3
B	Trench length, feet Number of trenches	63 2	63 2	63 2	63 2
C	Trench spacing, feet	15	15	15	15
D	Mound height, inches	12	12	12	12
E	Mound height, inches	12	17	20	25
F	Mound height, inches	9	9	9	9
G	Mound height, inches	12	12	12	12
H	Mound height, inches	18	18	18	18
I	Mound width, feet[a]	12	20	20	20
J	Mound width, feet[a]	12	8	8	8
K	Mound length, feet	10	10	10	10
L	Mound length, feet	62	62	62	62
P	Distribution pipe length, feet Distribution pipe diameter, inches Number of holes per distribution pipe[b] Hole spacing, inches[b] Hole diameter, inches	31 $1^1/_4$ 13 30 0.25	31 $1^1/_4$ 13 30 0.25	31 $1^1/_4$ 13 30 0.25	31 $1^1/_4$ 13 30 0.25
R	Manifold length, feet Manifold diameter, inches[c]	15 2	15 2	15 2	15 2
W	Mound width, feet	42	46	46	46

For SI: 1 inch = 25.4 mm, 1 foot = 304.8 mm, 1 gallon = 3.785 L.

a. Additional width to obtain required basal area.

b. First hole is located 12 inches from the manifold.

c. Diameter dependent on the size of pipe from pump and inlet position.

TABLE 903.1(4)—DESIGN CRITERIA FOR A FOUR-BEDROOM HOME FOR A MOUND ON A 0- TO 6-PERCENT SLOPE WITH LOADING RATES OF 600 GALLONS PER DAY FOR SLOWLY PERMEABLE SOIL

DESIGN PARAMETER		SLOPE (percent)			
		0	2	4	6
A	Trench width, feet	3	3	3	3
B	Trench length, feet Number of trenches	563	563	563	563
C	Trench spacing, feet	15	15	15	15
D	Mound height, inches	12	12	12	12
E	Mound height, inches	12	20	28	36
F	Mound height, inches	9	9	9	9
G	Mound height, inches	12	12	12	12
H	Mound height, inches	24	24	24	24
I	Mound width, feet[a]	12	20	20	20
J	Mound width, feet[a]	12	8	8	8
K	Mound length, feet	12	12	12	14
L	Mound length, feet	80	80	80	84
P	Distribution pipe length, feet Distribution pipe diameter, inches Number of holes per distribution pipe[b] Hole spacing, inches[b] Hole diameter, inches	27.5 $1^1/_4$ 12 30 0.25	27.5 $1^1/_4$ 12 30 0.25	27.5 $1^1/_4$ 12 30 0.25	27.5 $1^1/_4$ 12 30 0.25
R	Manifold length, feet Manifold diameter, inches[c]	30 2	30 2	30 2	30 2
W	Mound width, feet	57	61	61	61

For SI: 1 inch = 25.4 mm, 1 foot = 304.8 mm, 1 gallon = 3.785 L.

a. Additional width to obtain required basal area.

b. Last hole is located at the end of the distribution pipe, which is 15 inches from the previous hole.

c. Diameter dependent on the size of pipe from pump and inlet position.

TABLE 903.1(5)—DESIGN CRITERIA FOR A ONE-BEDROOM HOME FOR A MOUND ON A 0- TO 12-PERCENT SLOPE WITH LOADING RATES OF 150 GALLONS PER DAY FOR SHALLOW PERMEABLE SOIL OVER CREVICED BEDROCK

DESIGN PARAMETER		PERCOLATION RATE (minutes per inch) SLOPE (percent)						
		3 to 60				3 to less than 30		
		0	2	4	6	8	10[a]	12[a]
A	Bed width, feet[b]	10	10	10	10	10	10	10
B	Bed length, feet	13	13	13	13	13	13	13
D	Mound height, inches	24	24	24	24	24	24	24
E	Mound height, inches	24	26	29	31	34	36	38
F	Mound height, inches	9	9	9	9	9	9	9
G	Mound height, inches	12	12	12	12	12	12	12
H	Mound height, inches	18	18	18	18	18	18	18
I	Mound width, feet	12	13	14	17	18	21	26
J	Mound width, feet	12	11	10	10	9	9	9
K	Mound length, feet	12	12	12	13	13	13	15
L	Mound length, feet	37	37	37	39	39	39	43
P	Distribution pipe length, feet[c] Distribution pipe diameter, inches Number of distribution pipes	12.5 1 6	12.5 1 6	12.5 1 6	12.5 1 6	12.5 1 6	12.5 1 6	12.5 1 6
R	Manifold length, feet Manifold diameter, inches[c]	6 2	6 2	6 2	6 2	6 2	6 2	6 2

TABLE 903.1(5)—DESIGN CRITERIA FOR A ONE-BEDROOM HOME FOR A MOUND ON A 0- TO 12-PERCENT SLOPE WITH LOADING RATES OF 150 GALLONS PER DAY FOR SHALLOW PERMEABLE SOIL OVER CREVICED BEDROCK—continued

DESIGN PARAMETER		PERCOLATION RATE (minutes per inch) SLOPE (percent)						
		3 to 60				3 to less than 30		
		0	2	4	6	8	10[a]	12[a]
S	Distribution pipe spacing, feet Number of holes per distribution pipe[d] Hole spacing, inches[d] Hole diameter, inches	3 6 30 0.25	3 6 30 0.25	3 6 30 0.25	3 6 30 0.25	3 6 30 0.25	3 6 30 0.25	3 6 30 0.25
W	Mound width, feet	34	34	34	37	37	41	45

For SI: 1 inch = 25.4 mm, 1 foot = 304.8 mm, 1 gallon = 3.785 L, 1 minute per inch = 2.4 s/mm.

a. On sites with a 10- to 12-percent slope, the fill depth (*D*) shall be reduced to not less than 1.5 feet or the bed width shall be reduced to decrease *E* (downslope fill depth, feet).

b. Bed widths shall not be limited.

c. Use a manifold with distribution pipes on only one side.

d. Last hole is located at the end of the distribution pipe, which is 15 inches from the previous hole.

TABLE 903.1(6)—DESIGN CRITERIA FOR A TWO-BEDROOM HOME FOR A MOUND ON A 0- TO 12-PERCENT SLOPE WITH LOADING RATES OF 300 GALLONS PER DAY FOR SHALLOW PERMEABLE SOIL OVER CREVICED BEDROCK

DESIGN PARAMETER		PERCOLATION RATE (minutes per inch) SLOPE (percent)						
		3 to 60			3 to less than 30			
		0	2	4	6	8	10[a]	12[a]
A	Bed width, feet[b]	10	10	10	10	10	10	10
B	Bed length, feet	25	25	25	25	25	25	25
D	Mound height, inches	24	24	24	24	24	24	24
E	Mound height, inches	24	26	29	31	34	36	38
F	Mound height, inches	9	9	9	9	9	9	9
G	Mound height, inches	12	12	12	12	12	12	12
H	Mound height, inches	18	18	18	18	18	18	18
I	Mound width, feet	12	13	14	17	18	21	26
J	Mound width, feet	12	11	10	10	9	9	9
K	Mound length, feet	12	12	12	13	13	13	15
L	Mound length, feet	49	49	49	51	51	51	55
P	Distribution pipe length, feet[c] Distribution pipe diameter, inches Number of distribution pipes	12 1 6	12 1 6	12 1 6	12 1 6	12 1 6	12 1 6	12 1 6
R	Manifold length, feet Manifold diameter, inches	6 2	6 2	6 2	6 2	6 2	6 2	6 2
S	Distribution pipe spacing, feet Number of holes per distribution pipe[d] Hole spacing, inches[d] Hole diameter, inches	3 5 30 0.25	3 5 30 0.25	3 5 30 0.25	3 5 30 0.25	3 5 30 0.25	3 5 30 0.25	3 5 30 0.25
W	Mound width, feet	34	34	34	37	37	41	45

For SI: 1 inch = 25.4 mm, 1 foot = 304.8 mm, 1 gallon = 3.785 L, 1 minute per inch = 2.4 s/mm.

a. On sites with a 10- to 12-percent slope, the fill depth (*D*) shall be reduced to not less than 1.5 feet or the bed width shall be reduced to decrease *E* (downslope fill depth, feet).

b. Bed widths shall not be limited.

c. This design is based on a manifold with distribution pipes on both sides. An alternative design basis is 24-foot distribution pipes, with manifold at the end.

d. Last hole is located 9 inches from the end of the distribution pipe.

TABLE 903.1(7)—DESIGN CRITERIA FOR A THREE-BEDROOM HOME FOR A MOUND ON A 0- TO 12-PERCENT SLOPE WITH LOADING RATES OF 450 GALLONS PER DAY FOR SHALLOW PERMEABLE SOIL OVER CREVICED BEDROCK

DESIGN PARAMETER		PERCOLATION RATE (minutes per inch) SLOPE (percent)						
		3 to 60		3 to less than 30				
		0	2	4	6	8	10[a]	12[a]
A	Bed width, feet[b]	10	10	10	10	10	10	10
B	Bed length, feet	38	38	38	38	38	38	38
D	Mound height, inches	24	24	24	24	24	24	24
E	Mound height, inches	24	26	29	31	34	36	38
F	Mound height, inches	9	9	9	9	9	9	9
G	Mound height, inches	12	12	12	12	12	12	12
H	Mound height, inches	18	18	18	18	18	18	18
I	Mound width, feet	12	13	14	17	18	21	26
J	Mound width, feet	12	11	10	10	9	9	9
K	Mound length, feet	12	12	12	13	13	13	15
L	Mound length, feet	62	62	62	64	64	64	68
P	Distribution pipe length, feet[c] Distribution pipe diameter, inches Number of distribution pipes	18.5 1 6	18.5 1 6	18.5 1 6	18.5 1 6	18.5 1 6	18.5 1 6	18.5 1 6
R	Manifold length, feet Manifold diameter, inches	6 2	6 2	6 2	6 2	6 2	6 2	6 2
S	Distribution pipe spacing, feet Number of holes per distribution pipe[d] Hole spacing, inches[d] Hole diameter, inches	3 8 30 0.25	3 8 30 0.25	3 8 30 0.25	3 8 30 0.25	3 8 30 0.25	3 8 30 0.25	3 8 30 0.25
W	Mound width, feet	34	34	34	37	37	41	45

For SI: 1 inch = 25.4 mm, 1 foot = 304.8 mm, 1 gallon = 3.785 L, 1 minute per inch = 2.4 s/mm.

a. On sites with a 10- to 12-percent slope, the fill depth (*D*) shall be reduced to not less than 1.5 feet or the bed width shall be reduced to decrease *E* (downslope fill depth, feet).

b. Bed widths shall not be limited.

c. Use a manifold with distribution pipes on only one side.

d. Last hole is located at the end of the distribution pipe, which is 27 inches from the previous hole.

TABLE 903.1(8)—DESIGN CRITERIA FOR A FOUR-BEDROOM HOME FOR A MOUND ON A 0- TO 12-PERCENT SLOPE WITH LOADING RATES OF 600 GALLONS PER DAY FOR SHALLOW PERMEABLE SOIL OVER CREVICED BEDROCK

DESIGN PARAMETER		PERCOLATION RATE (minutes per inch) SLOPE (percent)						
		3 to 60				3 to less than 30		
		0	2	4	6	8	10[a]	12[a]
A	Bed width, feet[b]	10	10	10	10	10	10	10
B	Bed length, feet	50	50	50	50	50	50	50
D	Mound height, inches	24	24	24	24	24	24	24
E	Mound height, inches	24	26	29	31	34	36	38
F	Mound height, inches	9	9	9	9	9	9	9
G	Mound height, inches	12	12	12	12	12	12	12
H	Mound height, inches	18	18	18	18	18	18	18
I	Mound width, feet	12	13	14	17	18	21	26
J	Mound width, feet	12	11	10	10	9	9	9
K	Mound length, feet	12	12	12	13	13	13	15
L	Mound length, feet	74	74	74	76	76	76	78
P	Distribution pipe length, feet[c] Distribution pipe diameter, inches Number of distribution pipes	24.5 1 6	24.5 1 6	24.5 1 6	24.5 1 6	24.5 1 6	24.5 1 6	24.5 1 6

TABLE 903.1(8)—DESIGN CRITERIA FOR A FOUR-BEDROOM HOME FOR A MOUND ON A 0- TO 12-PERCENT SLOPE WITH LOADING RATES OF 600 GALLONS PER DAY FOR SHALLOW PERMEABLE SOIL OVER CREVICED BEDROCK—continued

DESIGN PARAMETER		PERCOLATION RATE (minutes per inch) SLOPE (percent)						
		3 to 60				3 to less than 30		
		0	2	4	6	8	10[a]	12[a]
R	Manifold length, feet Manifold diameter, inches	6 2	6 2	6 2	6 2	6 2	6 2	6 2
S	Distribution pipe spacing, feet Number of holes per distribution pipe[d] Hole spacing, inches[d] Hole diameter, inches	3 10 30 0.25	3 10 30 0.25	3 10 30 0.25	3 10 30 0.25	3 10 30 0.25	3 10 30 0.25	3 10 30 0.25
W	Mound width, feet	34	34	34	37	37	41	45

For SI: 1 inch = 25.4 mm, 1 foot = 304.8 mm, 1 gallon = 3.785 L, 1 minute per inch = 2.4 s/mm.
a. On sites with a 10- to 12-percent slope, the fill depth (*D*) shall be reduced to not less than 1.5 feet or the bed width shall be reduced to decrease *E* (downslope fill depth, feet).
b. Bed widths shall not be limited.
c. Use a manifold with distribution pipes on only one side.
d. Last hole is located 9 inches from the end of the distribution pipe.

TABLE 903.1(9)—DESIGN CRITERIA FOR A ONE-BEDROOM HOME FOR A MOUND ON A 0- TO 12-PERCENT SLOPE WITH LOADING RATES OF 150 GALLONS PER DAY FOR PERMEABLE SOIL WITH A HIGH WATER TABLE

DESIGN PARAMETER		PERCOLATION RATE (minutes per inch) SLOPE (percent)						
		0 to 60				0 to less than 30		
		0	2	4	6	8	10	12
A	Bed width, feet	4	4	4	4	4	4	4
B	Bed length, feet	32	32	32	32	32	32	32
D	Mound height, inches	12	12	12	12	12	12	12
E	Mound height, inches	12	13	14	14	16	17	18
F	Mound height, inches	9	9	9	9	9	9	9
G	Mound height, inches	12	12	12	12	12	12	12
H	Mound height, inches	18	18	18	18	18	18	18
I	Mound width, feet	9	10	11	12	13	14	15
J	Mound width, feet	9	9	8	8	7	7	6
K	Mound length, feet	10	10	10	10	10	11	11
L	Mound length, feet	52	52	52	52	52	53	53
P	Distribution pipe length Distribution pipe diameter, inches Number of distribution pipes Number of holes per distribution pipe[a] Hole spacing, inches[a] Hole diameter, inches	15.5 1 2 7 30 0.25	15.5 1 2 7 30 0.25	15.5 1 2 7 30 0.25	15.5 1 2 7 30 0.25	15.5 1 2 7 30 0.25	15.5 1 2 7 30 0.25	15.5 1 2 7 30 0.25
W	Mound width, feet	22	23	23	24	24	25	25

For SI: 1 inch = 25.4 mm, 1 foot = 304.8 mm, 1 gallon = 3.785 L, 1 minute per inch = 2.4 s/mm.
a. Last hole is located at the end of the distribution pipe, which is 21 inches from the previous hole.

TABLE 903.1(10)—DESIGN CRITERIA FOR A TWO-BEDROOM HOME FOR A MOUND ON A 0- TO 12-PERCENT SLOPE WITH LOADING RATES OF 300 GALLONS PER DAY FOR PERMEABLE SOIL WITH A HIGH WATER TABLE

DESIGN PARAMETER		PERCOLATION RATE (minutes per inch) SLOPE (percent)						
		0 to 60				0 to less than 30		
		0	2	4	6	8	10	12
A	Bed width, feet	6	6	6	6	6	6	6

TABLE 903.1(10)—DESIGN CRITERIA FOR A TWO-BEDROOM HOME FOR A MOUND ON A 0- TO 12-PERCENT SLOPE WITH LOADING RATES OF 300 GALLONS PER DAY FOR PERMEABLE SOIL WITH A HIGH WATER TABLE—continued

DESIGN PARAMETER		PERCOLATION RATE (minutes per inch) SLOPE (percent)						
		0 to 60				0 to less than 30		
		0	2	4	6	8	10	12
B	Bed length, feet	42	42	42	42	42	42	42
D	Mound height, inches	12	12	12	12	12	12	12
E	Mound height, inches	12	13	14	17	18	19	22
F	Mound height, inches	9	9	9	9	9	9	9
G	Mound height, inches	12	12	12	12	12	12	12
H	Mound height, inches	18	18	18	18	18	18	18
I	Mound width, feet	9	10	11	12	13	15	16
J	Mound width, feet	9	9	8	8	7	7	6
K	Mound length, feet	10	10	10	10	10	11	11
L	Mound length, feet	62	62	62	62	62	64	64
P	Distribution pipe length, feet[a] Distribution pipe diameter, inches Number of distribution pipes	20 1 4	20 1 4	20 1 4	20 1 4	20 1 4	20 1 4	20 1 4
R	Manifold length, feet Manifold diameter, inches	3 2	3 2	3 2	3 2	3 2	3 2	3 2
S	Distribution pipe spacing, feet Number of holes per distribution pipe[b] Hole spacing, inches[b] Hole diameter, inches	3 9 30 0.25	3 9 30 0.25	3 9 30 0.25	3 9 30 0.25	3 9 30 0.25	3 9 30 0.25	3 9 30 0.25
W	Mound width, feet	24	25	25	26	26	28	29

For SI: 1 inch = 25.4 mm, 1 foot = 304.8 mm, 1 gallon = 3.785 L, 1 minute per inch = 2.4 s/mm.

a. Use a manifold with distribution pipes only on one side.

b. Last hole is located at the end of the distribution pipe, which is 15 inches from the previous hole.

TABLE 903.1(11)—DESIGN CRITERIA FOR A THREE-BEDROOM HOME FOR A MOUND ON A 0- TO 12-PERCENT SLOPE WITH LOADING RATES OF 450 GALLONS PER DAY FOR PERMEABLE SOIL WITH A HIGH WATER TABLE

DESIGN PARAMETER		PERCOLATION RATE (minutes per inch) SLOPE (percent)						
		0 to 60				0 to less than 30		
		0	2	4	6	8	10	12
A	Bed width, feet	8	8	8	8	8	8	8
B	Bed length, feet	47	47	47	47	47	47	47
D	Mound height, inches	12	12	12	12	12	12	12
E	Mound height, inches	12	12	16	18	19	22	24
F	Mound height, inches	9	9	9	9	9	9	9
G	Mound height, inches	12	12	12	12	12	12	12
H	Mound height, inches	18	18	18	18	18	18	18
I	Mound width, feet	9	11	12	13	15	17	18
J	Mound width, feet	9	9	8	8	7	7	6
K	Mound length, feet	10	10	10	10	10	11	12
L	Mound length, feet	67	67	67	67	69	69	71
P	Distribution pipe length, feet Distribution pipe diameter, inches Number of distribution pipes	23 1 6	23 1 6	23 1 6	23 1 6	23 1 6	23 1 6	23 1 6
R	Manifold length, feet Manifold diameter, inches	64 2	64 2	64 2	64 2	64 2	64 2	64 2

TABLE 903.1(11)—DESIGN CRITERIA FOR A THREE-BEDROOM HOME FOR A MOUND ON A 0- TO 12-PERCENT SLOPE WITH LOADING RATES OF 450 GALLONS PER DAY FOR PERMEABLE SOIL WITH A HIGH WATER TABLE—continued

DESIGN PARAMETER		PERCOLATION RATE (minutes per inch) SLOPE (percent)						
		0 to 60				0 to less than 30		
		0	2	4	6	8	10	12
S	Distribution pipe spacing, feet Number of holes per distribution pipe[a] Hole spacing, inches[a] Hole diameter, inches	32 10 30 0.25	32 10 30 0.25	32 10 30 0.25	32 10 30 0.25	32 10 30 0.25	32 10 30 0.25	32 10 30 0.25
W	Mound width, feet	26	28	28	29	30	32	32

For SI: 1 inch = 25.4 mm, 1 foot = 304.8 mm, 1 gallon = 3.785 L, 1 minute per inch = 2.4 s/mm.
a. Last hole is located at the end of the distribution pipe, which is 21 inches from the previous hole.

TABLE 903.1(12)—DESIGN CRITERIA FOR A FOUR-BEDROOM HOME FOR A MOUND ON A 0- TO 12-PERCENT SLOPE WITH LOADING RATES OF 600 GALLONS PER DAY FOR PERMEABLE SOIL WITH A HIGH WATER TABLE

DESIGN PARAMETER		PERCOLATION RATE (minutes per inch) SLOPE (percent)						
		0 to 60				0 to less than 30		
		0	2	4	6	8	10	12
A	Bed width, feet	10	10	10	10	10	10	10
B	Bed length, feet	50	50	50	50	50	50	50
D	Mound height, inches	12	12	12	12	12	12	12
E	Mound height, inches	12	14	17	19	22	24	26
F	Mound height, inches	9	9	9	9	9	9	9
G	Mound height, inches	12	12	12	12	12	12	12
H	Mound height, inches	18	18	18	18	18	18	18
I	Mound width, feet	9	11	13	14	17	18	19
J	Mound width, feet	9	9	8	8	7	7	6
K	Mound length, feet	10	10	10	10	11	11	12
L	Mound length, feet	70	70	70	70	72	72	74
P	Distribution pipe length, feet Distribution pipe diameter, inches Number of distribution pipes	24.5 1 6	24.5 1 6	24.5 1 6	24.5 1 6	24.5 1 6	24.5 1 6	24.5 1 6
R	Manifold length, feet Manifold diameter, inches	6 2	6 2	6 2	6 2	6 2	6 2	6 2
S	Distribution pipe spacing, feet Number of holes per distribution pipe[a] Hole spacing, inches[a] Hole diameter, inches	3 10 30 0.25	3 10 30 0.25	3 10 30 0.25	3 10 30 0.25	3 10 30 0.25	3 10 30 0.25	3 10 30 0.25
W	Mound width, feet	28	29	31	32	34	35	36

For SI: 1 inch = 25.4 mm, 1 foot = 304.8 mm, 1 gallon = 3.785 L, 1 minute per inch = 2.4 s/mm.
a. Last hole is 9 inches from the end of the distribution pipe.

903.1.1 Symbols. The following symbols and notations shall apply to the provisions of this section.

A = Bed or trench width, feet (mm).

A_A = Required absorption area, square feet (m^2).

B = Bed or trench length, feet (mm).

B_A = Basal area, square feet (m^2).

C = Trench spacing, feet (mm).

C_I = Infiltration capacity of natural soil, gallons per foot per day (L/mm/day).

D = Fill depth, feet (mm).

E = Downslope fill depth, feet (mm).

F = Bed or trench depth, feet (mm).

G = Minimum cap and topsoil depth, feet (mm).

H = Cap and topsoil depth at center of mound, foot (mm).

I = Downslope width, feet (mm).

J = Upslope width, feet (mm).

K = End slope length, feet (mm).

L = Total mound length, feet (mm).

N = Number of trenches.

P = Distribution pipe length, feet (mm).

R = Manifold length, feet (mm).

S = Distribution pipe spacing, feet (mm).

S_D = Downslope correction factor.

S_U = Upslope correction factor.

T_W = Total daily wastewater flow, gallons per day (L/day).

W = Total mound width, feet (mm).

X = Slope, percent.

903.2 Size of absorption area. The absorption area shall be sized based on the daily wastewater flow and the infiltrative capacity of the medium sand texture fill material, equaling 1.2 gallons per square foot (0.03 L/m^2) per day. The required absorption area shall be determined by the following equation:

Equation 9-1 $$A_A = \frac{T_W}{1.2\ \text{gal./ft}^2\text{/day}}$$

For SI: 1 square foot = 0.0929 m^2, 1 gallon = 3.785 L.

903.3 Trenches. Effluent shall be distributed in the mound through a trench system for slowly permeable soils with or without high ground water. Trench length shall be selected by determining the longest dimension perpendicular to any slope on the site. Trench width and spacing is dependent on specific site conditions. Trenches shall be 2 feet to 4 feet (610 mm to 1219 mm) wide. Trench length (*B*) shall be not more than 100 feet (2540 mm). Trenches shall be of equal length where more than one trench is required. A mound shall not have more than three trenches. Trench spacing (*C*) shall be determined by the following equation:

Equation 9-2 $$C = \frac{T_W}{N \times 0.24\ \text{gal./ft}^2\text{/day} \times B}$$

For SI: 1 gallon = 3.785 L, 1 square foot = 0.0929 m^2.

The calculated trench spacing (*C*) shall be measured from center to center of the trenches. Facilities with more than 1,500 gallons (56 775 L) per day shall be specifically engineered and approved for use with a trench system.

903.4 Beds. A long, narrow bed design shall be used for permeable soils with high water tables. The bed shall be square or rectangular for shallow permeable soils over bedrock. The bed length (*B*) shall be set after determining the longest dimension available and perpendicular to any slope on the site.

903.5 Mound dimensions. The mound height consists of the fill depth, bed or trench depth, the cap and topsoil depth.

903.5.1 Fill depth. The fill depth (*D*) shall be not less than 1 foot (305 mm) for slowly permeable soils and permeable soils with high water tables and not less than 2 feet (610 mm) of fill shall be required for shallow permeable soils over bedrock. Additional fill shall be placed at the downslope end of the bed or trench where the site is not level so the bottom of the bed or trench is level. The downslope fill depth for bed systems shall be determined by the following equation:

Equation 9-3 $E = D + XA$

For SI: 1 foot = 304.8 mm.

The downslope fill depth for trench systems shall be determined by the following equation:

Equation 9-4 $E = D + X(C + A)$

For SI: 1 foot = 304.8 mm.

903.5.2 Bed or trench depth. The bed or trench depth (*F*) shall be not less than 9 inches (229 mm) and not less than 6 inches (152 mm) of aggregate shall be placed under the distribution pipes and not less than 2 inches (51 mm) of aggregate shall be placed over the top of the distribution pipes.

903.5.3 Cap and topsoil depth. The cap and topsoil depth (H) at the center of the mound shall be not less than 18 inches (457 mm), which includes 1 foot (305 mm) of subsoil and 6 inches (152 mm) of topsoil. Outer edges of the mound, G (the minimum cap and topsoil depth), shall be not less than 1 foot (305 mm), which includes 6 inches (152 mm) of subsoil and 6 inches (152 mm) of topsoil. The soil used for the cap shall be topsoil or finer textured subsoil.

903.5.4 Mound lengths. The total mound length (L) shall be determined by the following equation:

Equation 9-5 $L = B + 2K$

For SI: 1 foot = 304.8 mm.

where: $K = 3\left[\frac{(D+E)}{2} + F + H\right]$

903.5.5 Mound widths. The mound width for a bed system shall be determined by the following equation:

Equation 9-6 $W = J + A + I$

For SI: 1 foot = 304.8 mm.

The mound width for a trench system shall be determined by the following equation:

Equation 9-7 $W = J + \frac{A}{2} + C(N-1) - \frac{A}{2} + I$

where:

$J = 3(D + F + G)S_U$

$I = 3(E + F + G)S_D$

For SI: 1 foot = 304.8 mm.

The upslope correction factor (S_U) and the downslope correction factor (S_D) shall be determined based on the slope in accordance with Table 903.5.5.

TABLE 903.5.5—DOWNSLOPE AND UPSLOPE WIDTH CORRECTIONS FOR MOUNDS ON SLOPING SITES

SLOPE (percent)	DOWNSLOPE CORRECTION FACTOR (S_D)	UPSLOPE CORRECTION FACTOR (S_U)
0	1	1
1	1.03	0.97
2	1.06	0.94
3	1.10	0.915
4	1.14	0.89
5	1.18	0.875
6	1.22	0.86
7	1.27	0.83
8	1.32	0.80
9	1.38	0.785
10	1.44	0.77
11	1.51	0.75
12	1.57	0.73

903.6 Basal area. The minimum basal area required shall be determined by the following equation:

Equation 9-8 $B_A = \frac{T_W}{C_I}$

For SI: 1 square foot = 0.0929 m^2.

The infiltrative capacity of natural soil shall be determined on the percolation rate in accordance with Table 903.6.

TABLE 903.6—INFILTRATIVE CAPACITY OF NATURAL SOIL

PERCOLATION RATE (minutes per inch)	INFILTRATIVE CAPACITY (gallons per foot per day)
Less than 30	1.2
30 to 60	0.74
More than 60 to 120	0.24

For SI: 1 gallon per foot per day = 0.012 L/mm/day,1 minute per inch = 2.4 s/mm.

903.6.1 Basal area available in bed system. The available basal area for a bed system shall be determined by one of the following equations:

Equation 9-9 $B_A = B(A + I)$ for sloping sites

Equation 9-10 $B_A = BW$ for level sites

For SI: 1 square foot = 0.0929 m^2.

903.6.2 Basal area available in trench system. The available basal area for a trench system shall be determined by one of the following equations:

Equation 9-11 $$B_A = B\left(W + J + \frac{A}{2}\right)$$

Equation 9-12 $B_A = BW$ for level sites

For SI: 1 square foot = 0.0929 m^2.

903.6.3 Adequacy of basal area. The downslope width (I) on a sloping site shall be increased or the upslope width (J) and downslope (I) widths on a level site shall be increased until sufficient area is available if the basal area available is not equal to or greater than the basal area required.

903.7 Dose volume and pump. The dose volume and pump shall conform to the requirements of Chapters 7 and 8.

SECTION 904—CONSTRUCTION TECHNIQUES

904.1 General. Construction shall not commence where the soil is so wet a soil wire forms when the soil is rolled between the hands. Installation of mound systems where the soil on the site is frozen shall be prohibited for new construction.

904.2 Site preparation. Excess vegetation shall be cut and removed from the mound area. Small trees shall be cut to grade surface, leaving the stumps in place.

904.3 Force main. The force main from the pumping chamber shall be installed before the mound site is plowed. The force main shall be sloped uniformly toward the pumping chamber so the force main drains after each dose.

904.4 Plowing. The site shall be plowed with a moldboard plow or chisel plow. The site shall be plowed to a depth of 7 inches to 8 inches (178 mm to 203 mm) with the plowing perpendicular to the slope. Rototillers shall not be used. The sand fill shall be placed immediately after plowing. Foot and vehicular traffic shall be kept off the plowed area.

904.5 Sand fill material. The fill material shall be medium sand texture defined as 25 percent or more very coarse, coarse and medium sand and not more than 50 percent fine sand, very fine sand, silt and clay. The percentage of silt plus one and one-half times the percentage of clay shall not exceed 15 percent. Fill materials with higher content of silt and clay shall not be used.

904.5.1 Placement of sand fill. The medium sand fill shall be moved into place from the upslope and side edges of the plowed area. Vehicular traffic shall be prohibited in the area extending to 25 feet (7620 mm) beyond the downslope edge of the mound. The sand fill shall be moved into place with a track-type tractor and not less than 6 inches (152 mm) of sand shall be kept beneath the tracks at all times.

904.6 Installation of the absorption area. The bed or trenches shall be formed within the sand fill. The bottom of the trenches or bed shall be level. The elevation of the bottom of the trenches or bed shall be checked at the upslope and downslope edges to ensure that the fill has been placed to the proper depth.

904.7 Placement of the aggregate. Not less than 6 inches (152 mm) of coarse aggregate ranging in size from $^1/_2$ inch to $2^1/_2$ inches (12.7 mm to 64 mm) shall be placed in the bed or trench excavation. The top of the aggregate shall be level.

904.8 Distribution system. Distribution systems shall be placed on the aggregate, with the holes located on the bottom of the distribution pipe. The ends of all distribution pipes shall be marked at the surface, and an observation pipe shall be placed to the bottom of the bed or each trench.

904.9 Cover. The top of the bed or trenches shall be covered with not less than 2 inches (51 mm) of aggregate ranging in size from $^1/_2$ inch to $2^1/_2$ inches (12.7 mm to 64 mm) and not less than 4 inches to 5 inches (102 mm to 127 mm) of uncompacted straw or marsh hay or approved synthetic fabric shall be placed over the aggregate. Cap and topsoil covers shall be in place and the mound shall be seeded immediately and protected from erosion.

904.10 Maintenance. When the septic tank is pumped, the pump chamber shall be inspected and pumped to remove any solids present. Excess traffic in the mound area shall be avoided.

CESSPOOLS

User notes:

About this chapter: *Chapter 10 addresses the construction of cesspools. Although cesspools were a common method for disposing of sewage for new buildings in some areas of the United States, the code only allows the construction of a cesspool as a temporary measure until a proper sewage disposal system can be installed.*

SECTION 1001—GENERAL

1001.1 Scope. The provisions of this chapter shall govern the design and installation of *cesspools*.

1001.2 Application. *Cesspools* shall not be installed, except where approved by the *code official*. A *cesspool* shall be considered as only a temporary expedient pending the construction of a public sewer; as an overflow facility where installed in conjunction with an existing *cesspool*; or as a means of sewage disposal for limited, minor or temporary applications.

1001.3 Construction. *Cesspools* shall conform to the construction requirements of Section 605.3 for *seepage pits*. The *seepage pit* shall have a minimum sidewall of 20 feet (6096 mm) below the inlet opening. Where a stratum of gravel or equally pervious material of 4 feet (1219 mm) or more in thickness is found, the sidewall need not be more than 10 feet (3048 mm) below the inlet.

RESIDENTIAL WASTEWATER SYSTEMS

User notes:

About this chapter: *Septic tanks are not the only method for treatment of sewage from a residence. Chapter 11 indicates the standard with which factory-built wastewater treatment plants must comply.*

SECTION 1101—GENERAL

1101.1 Scope. The provisions of this chapter shall govern residential wastewater systems.

1101.2 Residential wastewater treatment systems. The regulations for materials, design, construction and performance shall comply with NSF 40 or IAPMO/ISO 30500.

INSPECTIONS

User notes:

About this chapter: *Chapter 12 covers the inspection requirements for all types of private sewage disposal systems.*

SECTION 1201—GENERAL

1201.1 Scope. The provisions of this chapter shall govern the inspection of *private sewage disposal systems.*

SECTION 1202—INSPECTIONS

1202.1 Initial inspection procedures. *Private sewage disposal systems* shall be inspected after construction, but before backfilling. The *code official* shall be notified when the *private sewage disposal system* is ready for inspection.

1202.2 Preparation for inspection. The installer shall make such arrangements as will enable the *code official* to inspect all parts of the system when a *private sewage disposal system* is ready. The installer shall provide the proper apparatus and equipment for conducting the inspection and furnish such assistance as is necessary to conduct the inspection.

1202.3 Covering of work. A *private sewage disposal system* or part thereof shall not be backfilled until such system has been inspected and approved. Any system that has been covered before being inspected and approved shall be uncovered as required by the *code official.*

1202.4 Other inspections. In addition to the required inspection prior to backfilling, the *code official* shall conduct any other inspections deemed necessary to determine compliance with this code.

1202.5 Inspections for additions, alterations or modifications. Additions, alterations or modifications to *private sewage disposal systems* shall be inspected.

1202.6 Defects in materials and workmanship. Where inspection discloses defective material, design or siting or unworkmanlike construction not conforming to the requirements of this code, the nonconforming parts shall be removed, replaced and reinspected.

CHAPTER 13 NONLIQUID SATURATED TREATMENT SYSTEMS

User notes:

About this chapter: *Composting is another method for disposing of human waste. Chapter 13 references the standard that covers these nonliquid saturated treatment systems.*

SECTION 1301—GENERAL

1301.1 Scope. The provisions of this chapter shall govern nonliquid saturated treament systems.

1301.2 Nonliquid saturated treatment systems. The regulations for materials, design, construction and performance shall comply with NSF 41.

CHAPTER

14 REFERENCED STANDARDS

User notes:

About this chapter: *This code contains numerous references to standards promulgated by other organizations that are used to provide requirements for materials and methods of construction. Chapter 14 contains a comprehensive list of all standards that are referenced in this code. These standards, in essence, are part of this code to the extent of the reference to the standard.*

ASTM

ASTM International, 100 Barr Harbor Drive, P.O. Box C700, West Conshohocken, PA 19428-2959

A74—2021: Specification for Cast Iron Soil Pipe and Fittings

Table 505.1

A888—21a: Specification for Hubless Cast Iron Soil Pipe and Fittings for Sanitary and Storm Drain, Waste, and Vent Piping Application

Table 505.1

B32—20: Specification for Solder Metal

505.8.2

B75/B75M—20: Specification for Seamless Copper Tube

Table 505.1

B88—20: Standard Specification for Seamless Copper Water Tube

Table 505.1

B251/B251M—2017: Specification for General Requirements for Wrought Seamless Copper and Copper-Alloy Tube

Table 505.1

B813—2016: Specification for Liquid and Paste Fluxes for Soldering of Copper and Copper Alloy Tube

505.8.2

B828—2016: Practice for Making Capillary Joints by Soldering of Copper and Copper Alloy Tube and Fittings

505.8.2

C4—2004(2018): Specification for Clay Drain Tile and Perforated Clay Drain Tile

Table 505.1

C14—20: Specification for Nonreinforced Concrete Sewer, Storm Drain and Culvert Pipe

Table 505.1

C76—22: Specification for Reinforced Concrete Culvert, Storm Drain and Sewer Pipe

Table 505.1

C425—21: Specification for Compression Joints for Vitrified Clay Pipe and Fittings

505.12, 505.13

C428/C428M—05(2011)e1: Specification for Asbestos-cement Nonpressure Sewer Pipe

Table 505.1

C443—21: Specification for Joints for Concrete Pipe and Manholes, Using Rubber Gaskets

505.7, 505.13

C478-15a: Specification for Circular Precast Reinforced Concrete Manhole Sections

504.1.1.2

C564—20a: Specification for Rubber Gaskets for Cast Iron Soil Pipe and Fittings

505.6.2, 505.6.3, 505.13

C700—2018: Specification for Vitrified Clay Pipe, Extra Strength, Standard Strength and Perforated

Table 505.1

C913—08: Specification for Precast Concrete Water and Wastewater Structures

504.1.1.3

C1173—2018: Specification for Flexible Transition Couplings for Underground Piping Systems

505.3.1, 505.5.1, 505.7, 505.10.1, 505.12, 505.13

C1277—20: Specification for Shielded Couplings Joining Hubless Cast Iron Soil Pipe and Fittings

504.1.1, 505.6.3

C1440—21: Specification for Thermoplastic Elastomeric (TPE) Gasket Materials for Drain, Waste and Vent (DWV), Sewer, Sanitary and Storm Plumbing Systems

505.13

C1460—21: Specification for Shielded Transition Couplings for Use with Dissimilar DWV Pipe and Fittings Above Ground
505.13

C1461—21: Specification for Mechanical Couplings Using Thermoplastic Elastomeric (TPE) Gaskets for Joining Drain, Waste and Vent (DWV) Sewer, Sanitary and Storm Plumbing Systems for Above and Below Ground Use
505.13

C1644—06: Specification for Resilient Connectors Between Reinforced Concrete On-Site Wastewater Tanks and Pipes
504.1.1

D1869—15: Specification for Rubber Rings for Fiber-Reinforced Cement Pipe
505.4, 505.13

D2235—2021: Specification for Solvent Cement for Acrylonitrile-Butadiene-Styrene (ABS) Plastic Pipe and Fittings
505.3.2, 505.5.2

D2564—2012(2018): Specification for Solvent Cements for Poly (Vinyl Chloride) (PVC) Plastic Piping Systems
505.10.2, 505.11.2

D2657—2007(2015): Standard Practice for Heat Fusion Joining of Polyolefin Pipe and Fittings
505.9.1

D2661—21: Specification for Acrylonitrile-butadiene-styrene (ABS) Schedule 40 Plastic Drain, Waste, and Vent Pipe and Fittings
Table 505.1, 505.3.2, 505.5.2

D2665—20: Specification for Poly (Vinyl Chloride) (PVC) Plastic Drain, Waste, and Vent Pipe and Fittings
Table 505.1

D2729—17: Specification for Poly (Vinyl Chloride) (PVC) Sewer Pipe and Fittings
Table 505.1.1

D2751—05: Specification for Acrylonitrile-Butadiene-Styrene (ABS) Sewer Pipe and Fittings
Table 505.1

D2855—2020: Standard Practice for the Two-Step (Primer and Solvent Cement) Method of Joining Poly (Vinyl Chloride) (PVC) or Chlorinated Poly (Vinyl Chloride) (CPVC) Pipe and Piping Components with Tapered Sockets
505.10.2, 505.11.2

D2949—18: Specification for 3.25-in. Outside Diameter Poly (Vinyl Chloride) (PVC) Plastic Drain, Waste, and Vent Pipe and Fittings
Table 505.1

D3034—2021: Specification for Type PSM Poly (Vinyl Chloride) (PVC) Sewer Pipe and Fittings
Table 505.1

D3212—2021: Specification for Joints for Drain and Sewer Plastic Pipes Using Flexible Elastomeric Seals
505.3.1, 505.5.1, 505.10.1

D4021—92: Specification for Glass-fiber Reinforced Polyester Underground Petroleum Storage Tanks
504.1.3

F405—05: Specification for Corrugated Polyethylene (PE) Pipe and Fittings
Table 505.1.1

F477—14(2021): Specification for Elastomeric Seals (Gaskets) for Joining Plastic Pipe
505.13

F628—2012E2: Specification for Acrylonitrile-Butadiene-Styrene (ABS) Schedule 40 Plastic Drain, Waste, and Vent Pipe with a Cellular Core
Table 505.1, 505.3.2, 505.5.2

F656—2015: Specification for Primers for Use in Solvent Cement Joints of Poly Vinyl Chloride (PVC) Plastic Pipe and Fittings
505.10.2, 505.11.2

F891—2016: Standard Specification for Coextruded Poly(Vinyl Chloride) (PVC) Plastic Pipe with a Cellular Core
Table 505.1

F1488—14(2019): Standard Specification for Coextruded Composite Pipe
Table 505.1, Table 505.1.1

F1499—2017: Specification for Coextruded Composite Drain Waste and Vent Pipe (DWV)
Table 505.1,

CISPI

Cast Iron Soil Pipe Institute, 2401 Fieldcrest Drive, Mundelein, IL 60060

301—21: Standard Specification for Hubless Cast Iron Soil Pipe and Fittings for Sanitary and Storm Drain, Waste and Vent Piping Applications
Table 505.1

310—20: Standard Specification for Coupling for Use in Connection with Hubless Cast Iron Soil Pipe and Fittings for Sanitary and Storm Drain, Waste and Vent Piping Applications
505.6.3

CSA *CSA Group, 8501 East Pleasant Valley Road, Cleveland, OH 44131-5516*

A257.2—19: Reinforced Circular Concrete Culvert, Storm Drain, Sewer Pipe and Fittings
Table 505.1

A257.3—19: Joints for Circular Concrete Sewer and Culvert Pipe, Manhole Sections and Fittings Using Rubber Gaskets
505.7, 505.13

B137.3—23: Rigid polyvinylchloride (PVC) Pipe and Fittings for Pressure Applications
505.10.2, 505.11.2

B181.1—21: Acrylonitrile-butadiene-styrene (ABS) Drain, Waste, and Vent Pipe and Pipe Fittings
505.3.2, 505.5.2

B181.2—21: Polyvinylchloride (PVC) and Chlorinated Polyvinylchloride (CPVC) Drain, Waste, and Vent Pipe and Pipe Fittings
505.10.2, 505.11.2

B182.1—21: Plastic Drain and Sewer Pipe and Pipe Fittings
505.10.2, 505.11.2

B182.2—21: PSM type Polyvinylchloride(PVC) Sewer Pipe and Fittings
Table 505.1

B182.4—21: Profile Polyvinylchloride (PVC) Sewer Pipe and Fittings
Table 505.1

B602—20: Mechanical Couplings for Drain, Waste, and Vent Pipe and Sewer Pipe
505.3.1, 505.5.1, 505.6.3, 505.7, 505.10.1, 505.12, 505.13

CSA A257.1—19: Non-reinforced Circular Concrete Culvert, Storm Drain, Sewer Pipe and Fittings
Table 505.1

ICC *International Code Council, Inc. 200 Massachusetts Avenue, NW, Suite 250, Washington, DC 20001*

IBC—24: International Building Code®
201.3

IPC—24: International Plumbing Code®
201.3, 505.14

ISO *International Organization for Standardization, Chemin de Blandonnet 8, Geneva, Switzerland 401-1214*

ANSI/CAN/IAPMO/ISO 30500-2019: Non-sewered sanitation systems — Prefabricated integrated treatment units — General safety and performance requirements for design and testing
1101.2

NSF *NSF International, 789 N. Dixboro Road, Ann Arbor, MI 48105*

40—2020: Residential Wastewater Treatment Systems
1101.2

41—2018: Nonliquid Saturated Treatment Systems (Composting Toilets)
1301.2

UL *UL LLC, 333 Pfingsten Road, Northbrook, IL 60062-2096*

70—2001: Septic Tanks, Bituminous-coated Metal
504.1.2

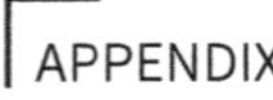

APPENDIX A

SYSTEM LAYOUT ILLUSTRATIONS

The provisions contained in this appendix are not mandatory unless specifically referenced in the adopting ordinance.

User notes:

About this chapter: *Appendix A provides illustrations for many system layouts covered in some of the previous chapters.*

Code development reminder: *Code change proposals to sections in this appendix will be considered by the IPC Code Development Committee during the 2024 (Group A) Code Development Cycle.*

SECTION A101—SYSTEM LAYOUTS

A101.1 System layouts. System layouts are illustrated in Figures A101.1(1) through A101.1(10).

FIGURE A101.1(1) (SECTION 403.1.1)—EXAMPLE OF SOIL-BORING LOCATIONS FOR TWO CONTIGUOUS ABSORPTION AREAS

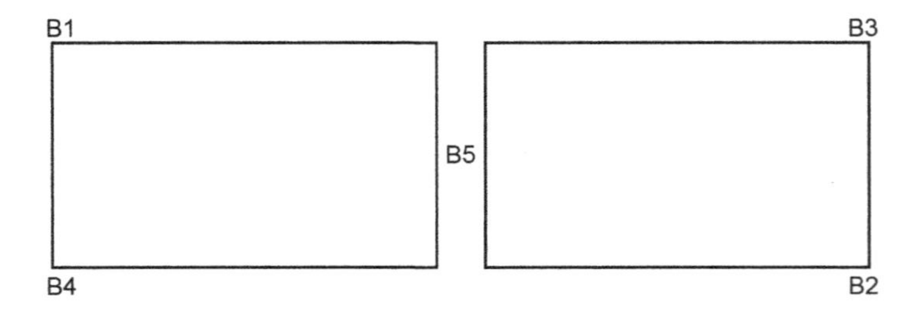

FIGURE A101.1(2) (SECTION 405.2.4)—MONITORING WELL DESIGN

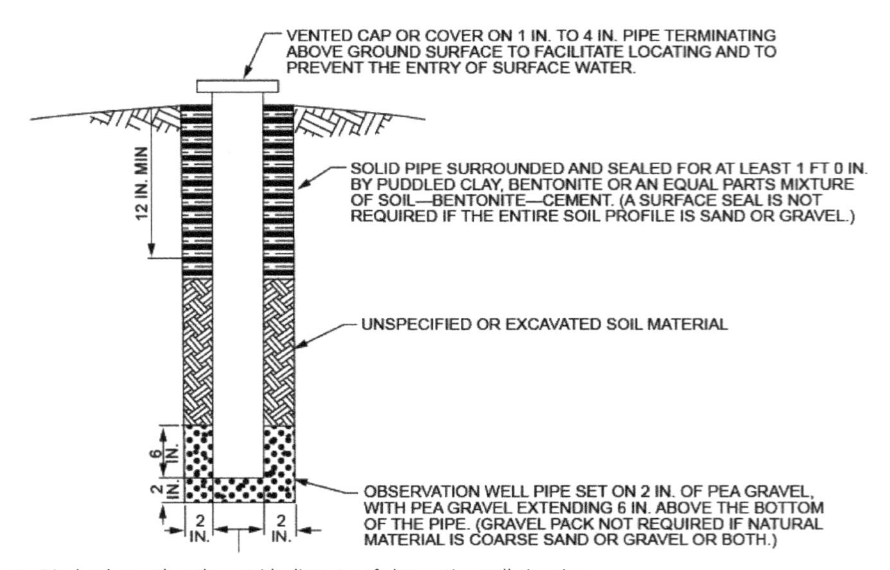

Note: Bore hole shall be 4 inches to 8 inches larger than the outside diameter of observation well pipe size.
For SI: 1 inch = 25.4 mm, 1 foot = 304.8 mm.

FIGURE A101.1(3) (SECTION 406.6.7)—DESIGN OF FILLED AREA SYSTEM

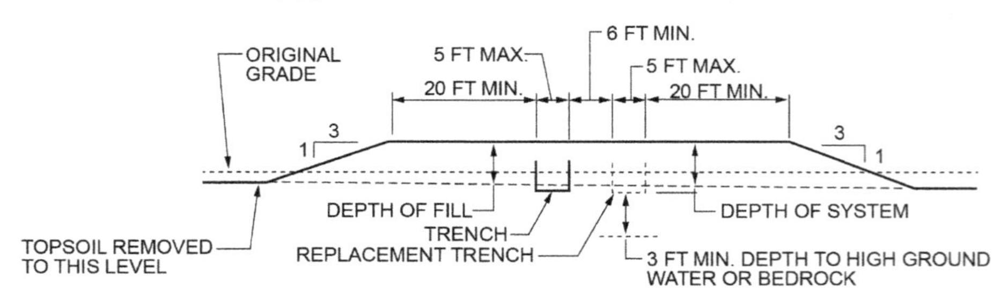

For SI: 1 foot = 304.8 mm.

FIGURE A101.1(4) (SECTION 605.7)—OBSERVATION PIPE

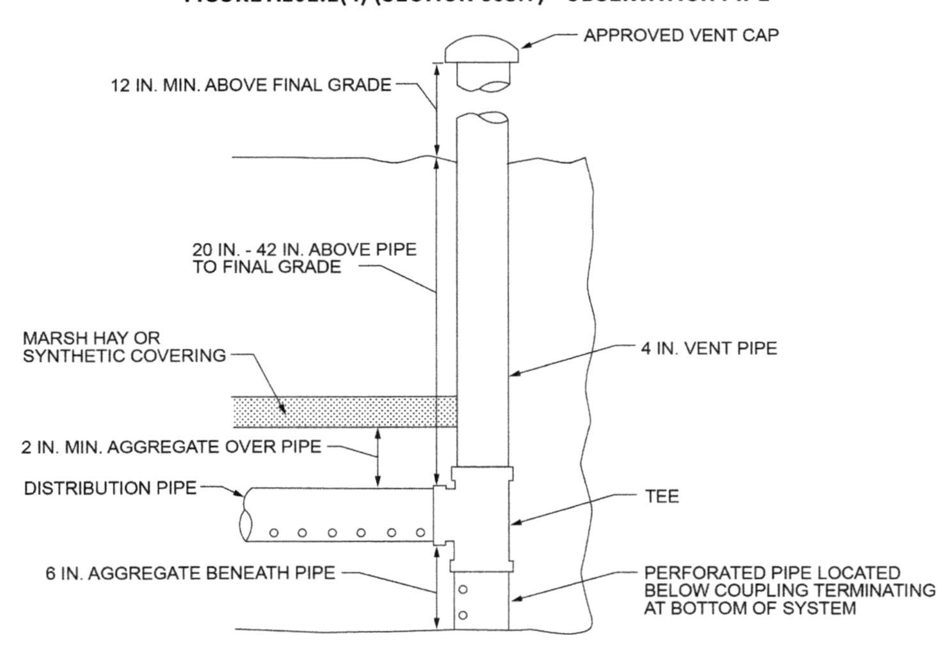

For SI: 1 inch = 25.4 mm.

FIGURE A101.1(5) (SECTION 703.1)—PRESSURE DISTRIBUTION SYSTEM DESIGN

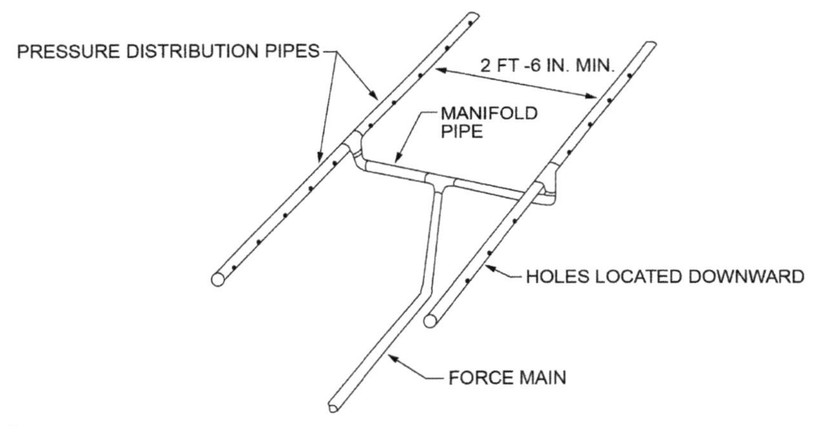

For SI: 1 inch = 25.4 mm, 1 foot = 304.8 mm.

FIGURE A101.1(6) (SECTION 903.1)—MOUND USING THREE TRENCHES FOR ABSORPTION AREA

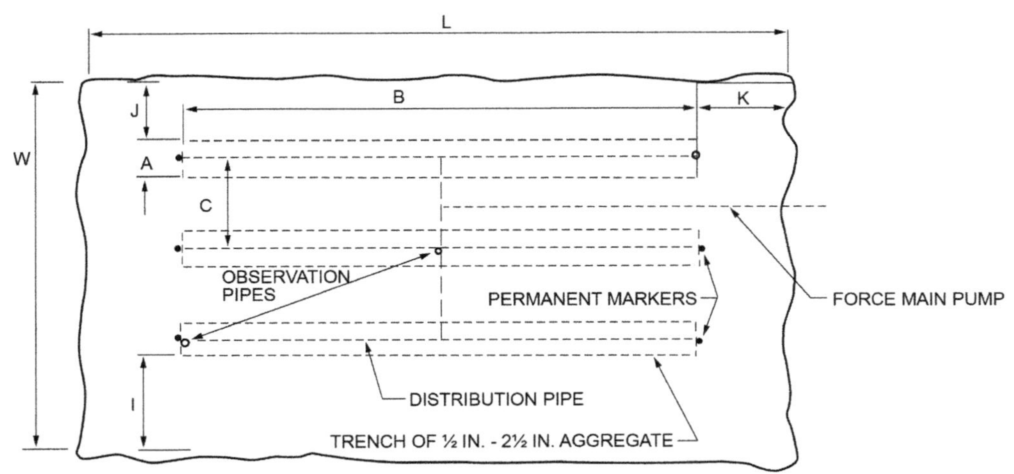

For SI: 1 inch = 25.4 mm.

FIGURE A101.1(7) (SECTION 903.1)—PLAN VIEW OF MOUND USING A BED FOR THE ABSORPTION AREA

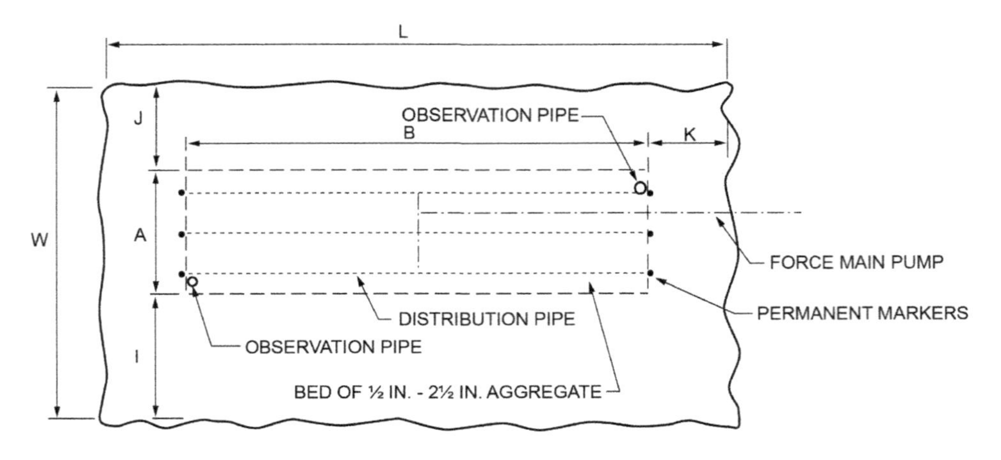

For SI: 1 inch = 25.4 mm.

FIGURE A101.1(8) (SECTION 903.1)—CROSS SECTION OF A MOUND SYSTEM USING THREE TRENCHES FOR THE ABSORPTION AREA

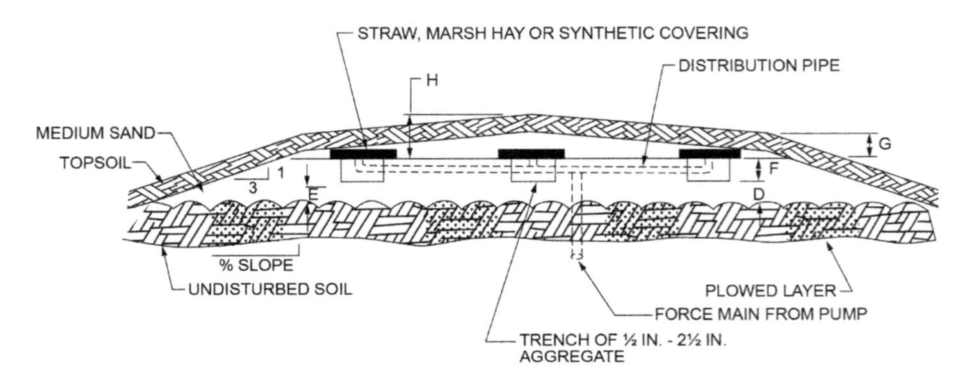

For SI: 1 inch = 25.4 mm.

FIGURE A101.1(9) (SECTION 903.1)—CROSS SECTION OF A MOUND SYSTEM USING A BED FOR THE ABSORPTION AREA

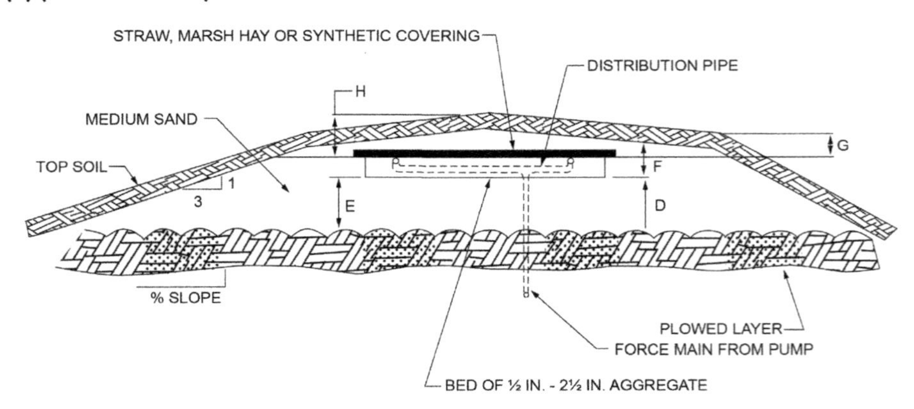

For SI: 1 inch = 25.4 mm.

FIGURE A101.1(10) (SECTION 903.1)—DISTRIBUTION PIPE LAYOUT

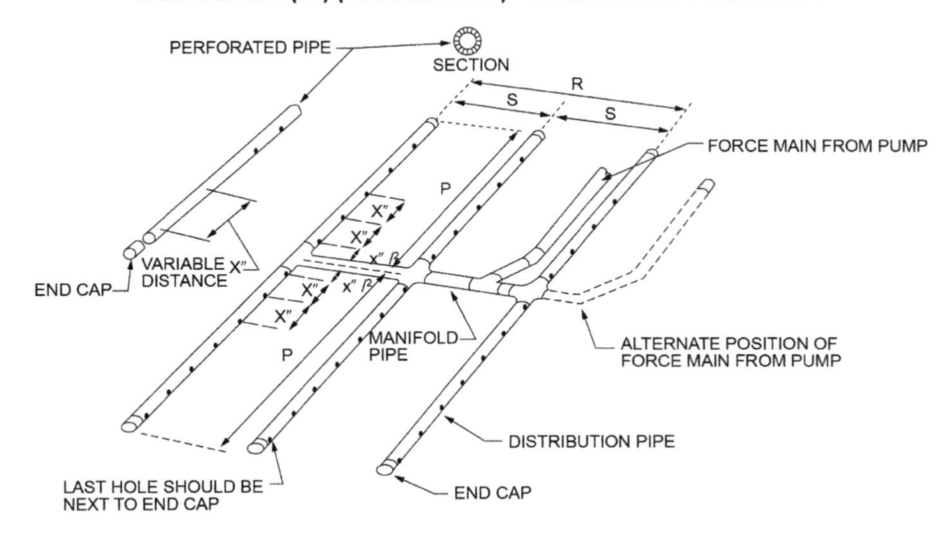

APPENDIX B

TABLES FOR PRESSURE DISTRIBUTION SYSTEMS

The provisions contained in this appendix are not mandatory unless specifically referenced in the adopting ordinance.

User notes:

About this chapter: *Appendix B provides design nomographs and tables for the design of pressure distribution systems.*

Code development reminder: *Code change proposals to sections in this appendix will be considered by the IPC Code Development Committee during the 2024 (Group A) Code Development Cycle.*

SECTION B101—PRESSURE DISTRIBUTION SYSTEMS

B101.1 General. The design of pressure distribution systems shall be in accordance with Tables B101.1(1) through B101.1(3) and Figures B101.1(1) through B101.1(3).

TABLE B101.1(1)—REQUIRED DISTRIBUTION PIPE DIAMETERS FOR VARIOUS HOLE DIAMETERS, HOLE SPACINGS AND DISTRIBUTION PIPE LENGTHS (SCHEDULE 40 PLASTIC PIPE)

DISTRIBUTION PIPE LENGTH (feet)	DISTRIBUTION PIPE DIAMETER (inch)																													
	Hole diameter (inch) 1/4						Hole diameter (inch) 5/16						Hole diameter (inch) 3/8						Hole diameter (inch) 7/16						Hole diameter (inch) 1/2					
	Hole spacing (feet)						Hole spacing (feet)						Hole spacing (feet)						Hole spacing (feet)						Hole spacing (feet)					
	2	3	4	5	6	7	2	3	4	5	6	7	2	3	4	5	6	7	2	3	4	5	6	7	2	3	4	5	6	7
10	1	1	1	1	1	1	1	1	1	1	1	1	1	1	1	1	1	1	1	1	1	1	1	1	1 1/4	1	1	1	1	1
15	1	1	1	1	1	1	1	1	1	1	1	1	1 1/4	1	1	1	1	1	1 1/4	1 1/4	1	1	1	1	1 1/4	1 1/4	1 1/4	1	1	1
20	1	1	1	1	1	1	1 1/4	1	1	1	1	1	1 1/4	1 1/4	1	1	1	1	1 1/4	1 1/4	1 1/4	1	1	1	2	1 1/2	1 1/4	1 1/4	1 1/4	1
25	1 1/4	1	1	1	1	1	1 1/4	1 1/4	1	1	1	1	1 1/2	1 1/4	1 1/4	1 1/4	1	1	2	1 1/2	1 1/4	1 1/4	1 1/4	1 1/4	2	2	1 1/2	1 1/4	1 1/4	1 1/4
30	1 1/4	1 1/4	1	1	1	1	1 1/2	1 1/4	1 1/4	1 1/4	1	1	2	1 1/2	1 1/2	1 1/4	1 1/4	1 1/4	2	2	1 1/2	1 1/4	1 1/4	1 1/4	3	2	2	1 1/2	1 1/2	1 1/4
35	1 1/2	1 1/4	1 1/4	1	1	1	2	1 1/2	1 1/4	1 1/4	1 1/4	1	2	2	1 1/2	1 1/4	1 1/4	1 1/4	3	2	2 1/2	1 1/2	1 1/2	1 1/4	3	3	2	2	1 1/2	1 1/2
40	1 1/2	1 1/4	1 1/4	1 1/4	1	1	2	1 1/2	1 1/2	1 1/4	1 1/4	1 1/4	3	2	1 1/2	1 1/2	1 1/4	1 1/4	3	2	2	2	1 1/2	1 1/2	3	3	2	2	2	1 1/2
45	2	1 1/2	1 1/4	1 1/4	1	1	2	2	1 1/2	1 1/4	1 1/4	1 1/4	3	2	2	1 1/2	1 1/2	1 1/2	3	3	2	2	2	1 1/2	3	3	3	2	2	2
50	2	1 1/2	1 1/4	1 1/4	1 1/4	1 1/4	3	2	2	1 1/2	1 1/2	1 1/4	3	3	2	2	2	1 1/2	3	3	2	2	2	1 1/2	3	3	3	3	2	2

For SI: 1 inch = 25.4 mm, 1 foot = 304.8 mm.

FIGURE B101.1(1)—DISTRIBUTION PIPE DISCHARGE RATE[a]

DISTRIBUTION PIPE OR MANIFOLD LENGTH (feet)

10 15 20 25 30 35 40 45 50

HOLE OR DISTRIBUTION PIPE SPACING (feet)

1 2 3 4 5 6 7

NUMBER OF HOLES

DISTRIBUTION PIPE DISCHARGE RATE (gallons per minute at 2½ feet head)

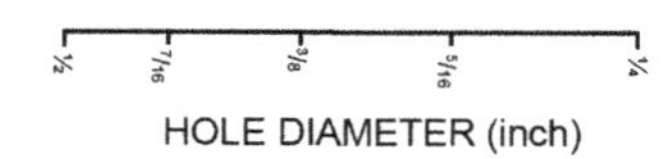

HOLE DIAMETER (inch)

For SI: 1 inch = 25.4 mm, 1 foot = 304.8 mm, 1 gallon per minute = 3.785 L/m.

a. Figure B101.1(1), a nomogram, determines the distribution pipe or manifold length, hole or distribution pipe spacing, number of holes, distribution discharge rate and hole diameter of pressure distribution systems by the placement of a straight edge between two known points.

TABLE B101.1(2)—RECOMMENDED MANIFOLD DIAMETERS FOR VARIOUS MANIFOLD LENGTHS, NUMBER OF DISTRIBUTION PIPES AND DISTRIBUTION PIPE DISCHARGE RATES (SCHEDULE 40 PLASTIC PIPE)

FLOW PER PIPE (gpm)	MANIFOLD LENGTH (feet) 5		10				15					20					25					30					FLOW PER PIPE (gpm)
	Number of distribution pipes with central manifold																										
	4	6	4	6	8	10	4	6	8	10	12	6	8	10	12	14	6	8	10	12	14	6	8	10	12	14	
	Manifold diameter (inch)																										
5	1	$1\frac{1}{4}$	$1\frac{1}{4}$	$1\frac{1}{4}$	$1\frac{1}{2}$	2	$1\frac{1}{4}$	$1\frac{1}{2}$	2	2	2	$1\frac{1}{4}$	$1\frac{1}{2}$	2	2	3	2	2	3	3	3	2	2	3	3	3	10
10	$1\frac{1}{4}$	$1\frac{1}{2}$	$1\frac{1}{2}$	2	2	3	2	2	3	3	3	2	3	3	3	3	3	3	3	3	3	3	3	3	4	4	20
15	$1\frac{1}{2}$	2	3	3	3	3	2	2	2	2	4	3	3	3	3	4	3	3	4	4	4	3	3	4	4	4	30
20	2	3	3	3	3	3	3	3	3	3	4	3	3	4	4	4	3	4	4	4	4	3	4	4	4	4	40
25	2	3	3	3	3	4	3	3	3	4	4	3	3	4	4	4	4	4	4	4	4	4	4	4	6	6	50
	Number of distribution pipes with end manifold																										
	2	3	2	3	4	5	2	3	4	5	6	3	4	5	6	7	3	4	5	6	7	3	4	5	6	7	

FLOW PER PIPE (gpm)	MANIFOLD LENGTH (feet) 35						40							45								50									FLOW PER PIPE (gpm)
	Number of distribution pipes with central manifold																														
	6	8	10	12	14	16	6	8	10	12	14	16	18	6	8	10	12	14	16	18	20	6	8	10	12	14	16	18	20	22	
	Manifold diameter (inch)																														
5	2	2	3	3	3	3	2	3	3	3	3	3	3	2	3	3	3	3	3	3	3	2	3	3	3	3	3	3	4	4	10
10	3	3	3	3	3	3	3	3	3	4	4	4	4	3	3	3	4	4	4	4	4	3	3	3	4	4	4	4	4	4	20
15	3	3	4	4	4	4	3	4	4	4	4	4	6	3	4	4	4	4	6	6	6	3	4	4	4	4	6	6	6	6	30
20	3	4	4	4	6	6	3	4	4	6	6	6	6	4	4	4	6	6	6	6	6	4	4	6	6	6	6	6	6	6	40
25	4	4	4	6	6	6	4	4	4	6	6	6	6	4	4	6	6	6	6	6	6	4	4	6	6	6	6	6	6	6	50
	Number of distribution pipes with end manifold																														
	3	4	5	6	7	8	3	4	5	6	7	8	9	3	4	5	6	7	8	9	10	3	4	5	6	7	8	9	10	11	

For SI: 1 inch = 25.4 mm, 1 foot = 304.8 mm, 1 gallon per minute = 3.785 L/m.

FIGURE B101.1(2)—PUMP DOSING RATE[a]

DISTRIBUTION PIPE DISCHARGE RATE (gallons per minute)

50 45 40 35 30 25 20 15 10 5 1

NUMBER OF DISTRIBUTION PIPES

1 5 10 15 20

DOSING RATE (gallons per minute)

1 10 20 30 40 50 100 150 200 300 400 500 1000

For SI: 1 gallon per minute = 3.785 L/m.

a. Figure B101.1(2), a nomogram, determines the distribution pipe or manifold length, hole or distribution pipe spacing, number of holes, distribution discharge rate and hole diameter of pressure distribution systems by the placement of a straight edge between two known points.

TABLE B101.1(3)—FRICTION LOSS[a] IN SCHEDULE 40 PLASTIC PIPE (C = 150)

FLOW (gpm)	PIPE DIAMETER (inch)								
	1	$1^1/_4$	$1^1/_2$	2	3	4	6	8	10
1	0.07	—	—	—	—	—	—	—	—
2	0.28	0.07	—	—	—	—	—	—	—
3	0.60	0.16	0.07	—	—	—	—	—	—
4	1.01	0.25	0.12	—	—	—	—	—	—
5	1.52	0.39	0.18	—	—	—	—	—	—
6	2.14	0.55	0.25	0.07	—	—	—	—	—
7	2.89	0.79	0.36	0.10	—	—	—	—	—
8	3.63	0.97	0.46	0.14	—	—	—	—	—
9	4.57	1.21	0.58	0.17	—	—	—	—	—
10	5.50	1.46	0.70	0.21	—	—	—	—	—
11	—	1.77	0.84	0.25	—	—	—	—	—
12	—	2.09	1.01	0.30	—	—	—	—	—
13	—	2.42	1.17	0.35	—	—	—	—	—
14	—	2.74	1.33	0.39	—	—	—	—	—
15	—	3.06	1.45	0.44	0.07	—	—	—	—
16	—	3.49	1.65	0.50	0.08	—	—	—	—
17	—	3.93	1.86	0.56	0.09	—	—	—	—
18	—	4.37	2.07	0.62	0.10	—	—	—	—
19	—	4.81	2.28	0.68	0.11	—	—	—	—
20	—	5.23	2.46	0.74	0.12	—	—	—	—
25	—	—	3.75	1.10	0.16	—	—	—	—
30	—	—	5.22	1.54	0.23	—	—	—	—
35	—	—	—	2.05	0.30	0.07	—	—	—
40	—	—	—	2.62	0.39	0.09	—	—	—
45	—	—	—	3.27	0.48	0.12	—	—	—
50	—	—	—	3.98	0.58	0.16	—	—	—
60	—	—	—	—	0.81	0.21	—	—	—
70	—	—	—	—	1.08	0.28	—	—	—
80	—	—	—	—	1.38	0.37	—	—	—
90	—	—	—	—	1.73	0.46	—	—	—
100	—	—	—	—	2.09	0.55	0.07	—	—
125	—	—	—	—	—	0.85	0.12	—	—
150	—	—	—	—	—	1.17	0.16	—	—
175	—	—	—	—	—	1.56	0.21	—	—
200	—	—	—	—	—	—	0.28	0.07	—

TABLE B101.1(3)—FRICTION LOSS[a] IN SCHEDULE 40 PLASTIC PIPE (C = 150)—continued

FLOW (gpm)	PIPE DIAMETER (inch)								
	1	$1^1/_4$	$1^1/_2$	2	3	4	6	8	10
250	Velocities in this area become too great for the various flow rates and pipe diameter.					—	0.41	0.11	—
300						—	0.58	0.16	—
350						—	0.78	0.20	0.07
400						—	0.99	0.26	0.09
450						—	1.22	0.32	0.11
500						—	—	0.38	0.14
600						—	—	0.54	0.18
700						—	—	0.72	0.24
800						—	—	—	0.32
900						—	—	—	0.38
1,000						—	—	—	0.46

For SI: 1 inch = 25.4 mm, 1 foot = 304.8 mm, 1 gallon per minute = 3.785 L/m.

a. Friction loss expressed in units of feet per 100 feet.

FIGURE B101.1(3)—MINIMUM DOSE VOLUME BASED ON PIPE SIZE, LENGTH AND NUMBER[a]

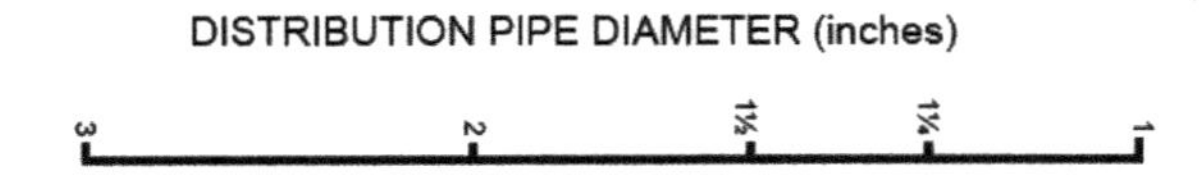

DISTRIBUTION PIPE LENGTH (feet)

10 15 20 25 30 35 40 45 50

PIPE VOLUME (gallons)

1 2 3 4 5 6 7 8 9 10 15 20

NUMBER OF DISTRIBUTION PIPES

20 15 10 5 4 3 2 1

DOSE VOLUME (gallons)

4000 3500 3000 2500 2000 1000 1000 900 800 700 600 500 450 400 350 300 250 200 150 100

For SI: 1 inch = 25.4 mm, 1 foot = 304.8 mm, 1 gallon = 3.787 L.

a. Figure B101.1(3), a nomogram, determines the distribution pipe or manifold length, hole or distribution pipe spacing, number of holes, distribution discharge rate and hole diameter of pressure distribution systems by the placement of a straight edge between two known points.

APPENDIX C

BOARD OF APPEALS

The provisions contained in this appendix are not mandatory unless specifically referenced in the adopting ordinance.

User Notes:

About this appendix: *Appendix C provides criteria for Board of Appeals members. Also provided are procedures by which the Board of Appeals should conduct its business.*

Code development reminder: *Code change proposals to this appendix will be considered by the Administrative Code Development Committee during the 2025 (Group B) Code Development Cycle.*

[A] SECTION C101—GENERAL

[A] C101.1 Scope. A board of appeals shall be established within the jurisdiction for the purpose of hearing applications for modification of the requirements of this code pursuant to the provisions of Section 112 (Means of Appeals). The board shall be established and operated in accordance with this section and shall be authorized to hear evidence from appellants and the *code official* pertaining to the application and intent of this code for the purpose of issuing orders pursuant to these provisions.

[A] C101.2 Application for appeal. Any person shall have the right to appeal a decision of the *code official* to the board. An application for appeal shall be based on a claim that the intent of this code or the rules legally adopted hereunder have been incorrectly interpreted, the provisions of this code do not fully apply or an equally good or better form of construction is proposed. The application shall be filed on a form obtained from the *code official* within 20 days after the notice was served.

[A] C101.2.1 Limitations of authority. The board shall not have authority to waive requirements of this code or interpret the administration of this code.

[A] C101.2.2 Stays of enforcement. Appeals of notice and orders, other than Imminent Danger notices, shall stay the enforcement of the notice and order until the appeal is heard by the board.

[A] C101.3 Membership of the board. The board shall consist of five voting members appointed by the chief appointing authority of the jurisdiction. Each member shall serve for **[NUMBER OF YEARS]** years or until a successor has been appointed. The board member's terms shall be staggered at intervals, so as to provide continuity. The *code official* shall be an ex officio member of said board but shall not vote on any matter before the board.

[A] C101.3.1 Qualifications. The board shall consist of five individuals who are qualified by experience and training to pass on matters pertaining to building construction and are not employees of the jurisdiction.

[A] C101.3.2 Alternate members. The chief appointing authority is authorized to appoint two alternate members who shall be called by the board chairperson to hear appeals during the absence or disqualification of a member. Alternate members shall possess the qualifications required for board membership, and shall be appointed for the same term or until a successor has been appointed.

[A] C101.3.3 Vacancies. Vacancies shall be filled for an unexpired term in the same manner in which original appointments are required to be made.

[A] C101.3.4 Chairperson. The board shall annually select one of its members to serve as chairperson.

[A] C101.3.5 Secretary. The chief appointing authority shall designate a qualified clerk to serve as secretary to the board. The secretary shall file a detailed record of all proceedings, which shall set forth the reasons for the board's decision, the vote of each member, the absence of a member and any failure of a member to vote.

[A] C101.3.6 Conflict of interest. A member with any personal, professional or financial interest in a matter before the board shall declare such interest and refrain from participating in discussions, deliberations and voting on such matters.

[A] C101.3.7 Compensation of members. Compensation of members shall be determined by law.

[A] C101.3.8 Removal from the board. A member shall be removed from the board prior to the end of their term only for cause. Any member with continued absence from regular meeting of the board may be removed at the discretion of the chief appointing authority.

[A] C101.4 Policies and procedures. The board shall establish policies and procedures necessary to carry out its duties consistent with the provisions of this code and applicable state law. The procedures shall not require compliance with strict rules of evidence but shall mandate that only relevant information be presented.

[A] C101.5 Notice of meeting. The board shall meet upon notice from the chairperson within 10 days of the filing of an appeal or at stated periodic intervals.

[A] C101.5.1 Open hearing. All hearings before the board shall be open to the public. The appellant, the appellant's representative, the *code official* and any person whose interests are affected shall be given an opportunity to be heard.

[A] C101.5.2 Quorum. Three members of the board shall constitute a quorum.

[A] C101.5.3 Postponed hearing. When five members are not present to hear an appeal, either the appellant or the appellant's representative shall have the right to request a postponement of the hearing.

[A] C101.6 Legal counsel. The jurisdiction shall furnish legal counsel to the board to provide members with general legal advice concerning matters before them for consideration. Members shall be represented by legal counsel at the jurisdiction's expense in all matters arising from service within the scope of their duties.

[A] C101.7 Board decision. The board shall only modify or reverse the decision of the *code official* by a concurring vote of three or more members.

[A] C101.7.1 Resolution. The decision of the board shall be by resolution. Every decision shall be promptly filed in writing in the office of the *code official* within 3 days and shall be open to the public for inspection. A certified copy shall be furnished to the appellant or the appellant's representative and to the *code official*.

[A] C101.7.2 Administration. The *code official* shall take immediate action in accordance with the decision of the board.

[A] C101.8 Court review. Any person, whether or not a previous party of the appeal, shall have the right to apply to the appropriate court for a writ of certiorari to correct errors of law. Application for review shall be made in the manner and time required by law following the filing of the decision in the office of the chief administrative officer.

INDEX

ICC Credentialing provides nationally recognized credentials that demonstrate a confirmed commitment to protect public health, safety, and welfare. Raise the professionalism of your department and further your career by pursuing an ICC Certification.

ICC Certifications Offer

- Nationwide recognition
- Increased earning potential
- Career advancement
- Superior knowledge
- Validation of your expertise
- Personal and professional satisfaction

Exams are developed and maintained to the highest standards, which includes continuous peer review by national committees of experienced, practicing professionals.

High Demand Exams (exam ID)

- Commercial Plumbing Inspector (P2)
- Plumbing Plans Examiner (P3)
- Commercial Mechanical Inspector (M2)
- Mechanical Plans Examiner (M3)
- Residential Building Inspector (B1)
- Commercial Building Inspector (B2)
- Building Plans Examiner (B3)

Proctored Remote Online Testing Option (PRONTO) provides a convenient testing experience that is accessible 24 hours a day, 7 days a week, 365 days a year.

Required hardware/ software is minimal – you will need a webcam and microphone, as well as a reasonably recent operating system.

Checkout all that ICC Credentialing has to offer at **iccsafe.org/certification**

Changes to the Code Development Process

The International Code Council is revising its code development process to improve the quality of code content by fostering a more in-depth vetting of code change proposals. Changes will take effect in 2024–2026 for the development of the 2027 I-Codes and will move the development process to an integrated and continuous three-year cycle.

In the new timeline:

- **Year 1** will include two Committee Action Hearings for Group A Codes
- **Year 2** will include two Committee Action Hearings for Group B Codes
- **Year 3** will be the joint Public Comment Hearings and Online Governmental Consensus Vote for both Group A and B Codes

Learn more about the new process and download the new timeline here.